NORTH CAROLINA
STATE BOARD OF COMMUNITY COLLEGES
LIBRARIES
ASHEVILLE-BUNCOMBE TECHNICAL COMMUNITY COLLEGE

DISCARDED

DEC - 6 2024

USE OF WASTE MATERIALS
IN HIGHWAY CONSTRUCTION

USE OF WASTE MATERIALS IN HIGHWAY CONSTRUCTION

by

Imtiaz Ahmed

Purdue University
Department of Civil Engineering
West Lafayette, Indiana

NOYES DATA CORPORATION
Park Ridge, New Jersey, U.S.A.

Copyright © 1993 by Noyes Data Corporation
Library of Congress Catalog Card Number: 92-541
ISBN: 0-8155-1315-1
Printed in the United States

Published in the United States of America by
Noyes Data Corporation
Mill Road, Park Ridge, New Jersey 07656

10 9 8 7 6 5 4 3 2 1

Library of Congress Cataloging-in-Publication Data

Ahmed, Imtiaz.
    Use of waste materials in highway construction / by Imtiaz Ahmed.
       p.   cm.
    Includes bibliographical references (p.) and index.
    ISBN 0-8155-1315-1
    1. Waste products as road materials. I. Title.
TE200.A45   1992
625.8--dc20   92-541
                                                                             CIP

Foreword

This book is an overview of the use of waste materials in highway construction. It presents a summary of current practices in the use of waste materials in highway construction and the experiences of many of the individual (U.S.) states in the technical, environmental, and economic aspects of the various applications of these waste materials.

The information presented was obtained from a review of published literature, recent unpublished reports, presentations of research updates by professionals at different forums, and personal meetings with experts. In addition, a questionnaire regarding the use of waste materials was sent to each U.S. state highway agency; and 42 agencies responded.

Public concern is constantly expressed about the vast quantities of useful materials being discarded or destroyed. Legislation to stimulate recycling efforts is in force in a number of states, and is being debated in others. This book presents one avenue of approach toward waste reduction and reuse.

The book describes the state-of-the-practice in the use of the waste materials in highway construction in the U.S. and discusses the applications of selected waste materials, including: waste tires, waste glass, reclaimed paving materials, slags and ashes, building rubble, sewage sludge, and incinerator residue. An evaluation based on technical, environmental, and economic factors indicated that reclaimed paving materials, coal fly ash, blast furnace slag, bottom ash, boiler slag, steel slag and rubber tires have significant potential to replace conventional materials for various applications in highway construction, and should be projected for future construction. Specific applications of the waste products and the potential problems associated with their usage in highway operations, which must be addressed prior to their extensive use, are also included.

The information in the book is from *Use of Waste Materials in Highway Construction,* prepared by Imtiaz Ahmed, Purdue University Department of Civil Engineering, for the Joint Highway Research Project of the Purdue University Engineering Experiment Station, in cooperation with the Indiana Department of Transportation and the U.S. Department of Transportation Federal Highway Administration, May 1991.

The table of contents is organized in such a way as to serve as a subject index and provides easy access to the information contained in the book.

> Advanced composition and production methods developed by Noyes Data Corporation are employed to bring this durably bound book to you in a minimum of time. Special techniques are used to close the gap between "manuscript" and "completed book." In order to keep the price of the book to a reasonable level, it has been partially reproduced by photo-offset directly from the original report and the cost saving passed on to the reader. Due to this method of publishing, certain portions of the book may be less legible than desired.

Acknowledgments

I wish to express my sincere thanks and profound sense of gratitude to Professor C.W. Lovell for his supervision and guidance during the course of this research. Thanks are also extended to Professor L.E. Wood who spared his much needed time for advice and also provided some of the references for this study.

I am thankful to the advisory committee members, especially Mr. D.W. Lucas, Ms. R.S. McDaniel, Mr. B.K. Partridge, and Mr. A.M. Rucker for their eagerness to help and also for sending the survey questionnaire to the United States highway agencies.

My thanks to all the United States highway agency officials who responded to the survey questionnaire of this study promptly. Recent research findings/performance evaluation studies provided by state highway agencies of California, Connecticut, Iowa, Kansas, Kentucky, Maine, Minnesota, Oregon, Texas, Vermont and Virginia are acknowledged with immense gratitude.

Financial support provided by the INDOT and the Federal Highway Administration through the Joint Highway Research Project, School of Civil Engineering, Purdue University, West Lafayette, Indiana, is appreciated.

Finally, I am grateful to my wife, Musarrat Imtiaz, for her continuous support and encouragement and my children for their patience during the course of this research.

Notice

The materials in this book were prepared as accounts of work sponsored by the Indiana Department of Transportation and the U.S. Department of Transportation Federal Highway Administration. On this basis the Publisher assumes no responsibility nor liability for errors or any consequences arising from the use of the information contained herein.

The contents of the book reflect the views of the author who is responsible for the facts and accuracy of the data presented. The contents do not necessarily reflect the official views or policies of the Federal Highway Administration, the Indiana DOT, or the Publisher. The book does not constitute a standard, specifications, or regulations.

Mention of trade names or commercial products does not constitute endorsement or recommendation for use by the Agencies or the Publisher. Final determination of the suitability of any information or product for use contemplated by any user, and the manner of that use, is the sole responsibility of the user. The book is intended for information purposes only. The reader is warned that caution must always be exercised regarding the use of waste materials, which could be potentially hazardous, and expert advice should be obtained before implementation.

Any information pertaining to law and regulations is provided for background only. The reader must contact the appropriate legal sources and regulatory authorities for up-to-date regulatory requirements, and their interpretation and implementation.

The book is sold with the understanding that the Publisher is not engaged in rendering legal, engineering, or other professional service. If advice or other expert assistance is required, the service of a competent professional should be sought.

Contents and Subject Index

1. **INTRODUCTION** 1
   1.1 **Background** 1
   1.2 **Objectives** 2
   1.3 **Research Approach** 2

2. **STATE-OF-THE-PRACTICE** 4
   2.1 **Indiana Department of Transportation** 4
      2.1.1 Introduction 4
      2.1.2 Operations/Maintenance 4
         Traffic Signs 4
         Delineators 4
         Used Oil 5
         Used Tires and Batteries 5
         Hazardous Waste Disposal 5
      2.1.3 Construction 6
         Reclaimed Asphalt Pavement 6
         Air Cooled Blast Furnace Slag 6
         Dry Bottom Ash 6
         Wet Bottom Ash (or Boiler Slag) 7
         Fly Ash 7
         Crumb Rubber 7
         Other Materials 7
      2.1.4 Research Into Future Uses 8
         The Role of INDOT in the Control of Hazardous
            Materials 8
         Asphalt Additive to Control Rutting and Cracking 8
         Evaluation of Crack Sealant Performance on

| | | Indiana's Bituminous Pavements .... 9 |
| | | Evaluation of Coal Bottom Ash .... 9 |

      2.1.5 Conclusions 9
  2.2 United States Highway Agencies 9
      2.2.1 Introduction 9
      2.2.2 An Overview of Current Practice 10

3. EVALUATION OF SELECTED WASTE PRODUCTS 23
  3.1 Rubber Tires 23
      3.1.1 Background 23
      3.1.2 Use of Scrap Tires in Asphalt Pavements 25
          Crack/Joint Sealant 26
          Surface/Interlayer Treatments 26
          Asphalt-Rubber Mixtures 27
          Rubber Modified Asphalt Mixtures 28
      3.1.3 Experience in the Use of Rubber Modified Asphalt
          Paving Products 29
          Alaska 29
          Arizona 30
          Connecticut 31
          Kansas 33
          Maine 33
          Minnesota 34
          New York 36
          Pennsylvania 37
          Vermont 38
          Washington 38
      3.1.4 Discussion 39
      3.1.5 Use of Tires in Subgrade/Embankment 41
          Use of Tires as Lightweight Aggregates 42
          Use of Tires for Soil Reinforcement 44
  3.2 Waste Glass 45
      3.2.1 Background 45
      3.2.2 Use of Glass in Asphalt Pavements 49
      3.2.3 Use of Waste Glass in Portland Cement Concrete 52
      3.2.4 Use of Glass in Unbound Aggregate Base Layers
          and Embankment Construction 52
      3.2.5 Discussion 53
  3.3 Reclaimed Paving Materials 55
      3.3.1 Recycling of Asphalt Pavements 56
          Iowa 57
          Kansas 58
      3.3.2 Recycling of Concrete Pavements 58
  3.4 Slags and Ashes 61

	3.4.1 Iron Blast Furnace Slag	61
	3.4.2 Steel Slag	63
	3.4.3 Bottom Ash	64
	3.4.4 Coal Fly Ash	66
	Use of Fly Ash in Cement Concrete Mixes	66
	Use of Fly Ash in Embankments	69
3.5	**Building Rubble**	71
3.6	**Sewage Sludge**	73
	3.6.1 Compost and Co-Compost	75
	3.6.2 Incinerator Ash from Sludge	76
3.7	**Incinerator Residue**	78

4.	SUMMARY, CONCLUSIONS AND RECOMMENDATIONS	82
4.1	Summary	82
4.2	Conclusions	83
	General	83
	Rubber Tires	83
	Waste Glass	85
	Reclaimed Paving Materials	86
	Slags and Ashes	86
	Building Rubble	87
	Sewage Sludge	88
	Incinerator Residue	88
	Other Materials	89
4.3	Recommendations	91
	Rubber Tires	91
	Waste Glass	92
	Reclaimed Paving Materials	92
	Slags and Ashes	92
	Building Rubble	93
	Sewage Sludge	93
	Incinerator Residue	93
	Other Materials	94

5.	REFERENCES	96
5.1	**References Cited**	96
5.2	**Other References**	106

APPENDIX A: SURVEY QUESTIONNAIRE	112
LIST OF ABBREVIATIONS	114

1. Introduction

1.1 Background

Enormous quantities of domestic, industrial and mining waste are generated annually in the United States. According to the Environmental Protection Agency (EPA, 1990), municipal solid waste (MSW) alone constituted 180 million tons in 1988. Without source reduction, generation of MSW is projected to reach 200 million tons by 1995. As the generation of waste continues to increase, the capacity to handle it is decreasing. Many landfills have closed, and new facilities are often difficult to site (EPA, 1990) due to economic and environmental constraints.

There are three techniques for waste disposal: (a) recycling; (b) incineration, with and without generation of energy; and (c) burial. The published data on current practice show that the bulk of domestic refuse is either incinerated or landfilled. Out of the total MSW generated in 1988: 13.1% was recovered; 14.2% was incinerated (13.6% with energy recovery and 0.6% without energy recovery); and 72.7% was landfilled. Public concern is constantly expressed about the vast quantities of useful materials being discarded or destroyed. Legislation which is intended to stimulate recycling efforts is in force in a number of states, and is being debated in others.

The Indiana Department of Transportation (INDOT) has been using recycled or waste products for many years in those applications which have been proven effective. They have also researched the use of a variety of waste products in highway construction to find an alternative source of material supply to offset the rising cost of quality natural aggregates, waste disposal, and energy. This study is also an attempt to search for additional waste products which are technically, economically, and environmentally feasible for use in highway construction by the INDOT.

1.2 Objectives

The principal objectives of this study were to:
- summarize the experience of INDOT in the use of waste materials in highway construction;
- determine what waste materials have been successfully used in other states and what applications have been proven effective in highway construction;
- recommend a plan to the INDOT for implementation of those waste products which have demonstrated technical, environmental and economic feasibility.

1.3 Research Approach

The tasks necessary to accomplish the stated objectives included:
- review of all available information on waste products use in highway construction;
- synthesis of the information that most fits the study objectives;
- reporting recommendations to the INDOT.

Two concurrent activities were pursued in conducting this study. The first activity consisted of a comprehensive literature review. Published material has been the main source of information. The following databases were searched to locate the literature on the subject:
- Compendex Plus (online form of engineering index);
- NTIS (National Technical Information Service);
- TRIS (Transportation Research Information System);
- Enviroline.
- Pollution Abstracts.

The formal literature search was supplemented by: recent unpublished reports/findings of research studies, presentations on research updates by professionals at different forums, and personal meetings with the experts.

The second activity consisted of a survey of waste materials and their current applications in the highway industry. A questionnaire regarding the use of waste materials was developed and distributed to each state highway agency. The questionnaire requested information on: the type of waste materials currently used in highway construction, their applications, annual quantities and field performance; the materials and applications which appeared favorable and would be projected for future construction; the materials and uses mandated by the state laws; and the waste materials that are used in a process covered by patents. In addition, the questionnaire requested copies of recent research findings/ performance evaluation studies on the use of waste materials in highway construction. A copy of the questionnaire is included in Appendix A.

The results of the questionnaire survey, literature review, and experience of the researchers were synthesized to prepare this report. This includes state-of-the-practice in the use of waste materials in highway construction in the United States and discussion on the applications of the selected waste products in the highway industry, considering technical feasibility, environmental consequences, and economic benefits. It also summarizes the conclusions of the study and recommends the materials, and their specific uses, which provide the best economic alternatives to conventional highway materials and which should be projected for future construction by the INDOT. Finally, areas that will require further research and testing are identified.

2. State-of-the-Practice

2.1 Indiana Department of Transportation

2.1.1 Introduction

This subsection briefly summarizes the experiences of the Indiana Department of Transportation (INDOT) in the use of recycled or waste products. It also provides estimates of the quantities of materials used and generated by the INDOT. The information contained in this subsection has been excerpted from Lucas (1990).

2.1.2 Operations/Maintenance

In the area of operation/maintenance activities, INDOT recycles some materials out of economic necessity and others because of the environmental laws. The following discusses the practices INDOT performs in various areas. These practices help to reduce INDOT's impact, directly and indirectly, on landfills.

Traffic Signs

INDOT has contract arrangements with Hall Contractors to refurbish up to 100,000 square feet of used aluminum traffic signs. Each district bundles old signs that have been damaged from accidents and sends them to the Traffic Annex where they are picked up by Hall to be refurbished. Therefore, these signs are not discarded but are used for new signs.

Delineators

Cast iron delineators or raised pavement markers are removed from the roads that are to be resurfaced. The recycling of the castings is written into the specifications of the resurfacing contracts. The highway contractor usually subcontracts this portion of the contract to a vendor. The refurbishing process entails removing the asphalt from the pavement marker and replacing the old reflective material with new. Again this demonstrates materials being reused. Thousands of delineator castings are reused every year.

Used Oil

INDOT generates approximately 55,000 gallons of waste oil annually. INDOT also generates small quantities of waste solvents from parts washers in the district garages. Some of the parts washers are serviced by companies such as Safety Kleen Oil Service, Buffalo, NY and Crystal Clean. In other instances, the waste solvents are combined with the used oil. The used oil is collected by various oil reclaimers in the state, who either charge for the collection or collect it without charge.

Used Tires and Batteries

Every year, the six districts of INDOT and the Toll Road have a surplus supplies sale. This is commonly known as a "210" lot sale. Used batteries, tires, used vehicles, scrap metal and other materials are sold there. This year (1990) INDOT sold over 6,000 used tires and approximately 850 used batteries.

INDOT is using a statewide quantity purchase order for batteries. The specifications incorporate the law that was passed earlier this year (1990) which requires the battery vendors to exchange one used battery for every new battery delivered. In addition, they pay $2 to INDOT for every used battery that is beyond the amount of new batteries delivered.

Hazardous Waste Disposal

INDOT has a hazardous waste disposal contract with Superior Oil, Inc., which collects INDOT's residual traffic paint materials, as well as the spent solvents used to clean painting equipment.

Superior Oil also picks up residual material from the INDOT Materials and Tests laboratories. This includes random mixtures of paints, alcohols, ketones, trichloroethane, mineral spirits, and toluene solutions. INDOT earlier used methylene chloride to extract asphalt from road paving mixtures for testing. Use of this hazardous solvent has been almost eliminated by switching to a biodegradable solvent, generically called limonene, made from orange peels. In 1989, 300 barrels of the solvents were used. One advantage of the use of the limonene is that there is no disposal problem; the solvent containing extracted asphalt is poured onto the aggregate stockpiles at a bituminous mixing plant and run back through the plant for incorporation into new bituminous road paving mixes.

2.1.3 Construction

Use of many types of waste products in highway construction is allowed by the INDOT Standard Specifications. These materials pose a disposal problem for those manufacturers who generate them. Their use as construction materials helps to reduce this disposal problem.

The actual quantities used are determined by the contractors, who usually select the materials based on economics. For instance, a concrete contractor may choose to replace part of the Portland cement in a mix with fly ash, if the ash is cheaper than cement. The following is a brief discussion of some of these materials and the quantities used.

<u>Reclaimed Asphalt Pavement</u>

The current INDOT Standard Specifications (INDOT, 1988) allow the use of up to 50% processed reclaimed asphalt pavement (RAP) from an approved source in new bituminous paving mixtures. On many contracts, the old surface must be milled off before the new materials are placed. Allowing this RAP to be reused reduces the amount of material going to landfills. Approximately 76,077 tons (38,420 cubic yards) of RAP were used in 1989.

<u>Air Cooled Blast Furnace Slag</u>

Air cooled blast furnace slag is a by-product of pig iron production in blast furnaces. This material is used as a replacement for naturally occurring aggregates. Approximately 2,000,000 tons of air cooled blast furnace slag were used by INDOT in bituminous and concrete construction in 1989. This tonnage represents over one million cubic yards of waste.

<u>Dry Bottom Ash</u>

These materials are produced from coal fired power plants and are heavy enough to drop to the bottom, falling through an open grate into a hopper. Bottom ashes have been used on one project in Indiana as an embankment material. Research conducted at Purdue University for INDOT indicates that some bottom ashes are suitable for use in embankments and fills if handled and enclosed properly (see Subsection 3.4.3). Ashes must be tested to determine if leachates are acceptable. Cost effectiveness of the use of bottom ash depends to a large extent on the location of a project with respect to the nearest acceptable source of bottom ash; trucking costs can greatly reduce the cost effectiveness.

Wet Bottom Ash (or Boiler Slag)

This is another waste material which is generated from coal fired power plants. It is produced when molten ash falls to the bottom of the boiler and is quenched in water. This type of slag is also used as an aggregate for bituminous construction (see Subsection 3.4.3). Approximately 100,000 tons (over 50,000 cubic yards) were used in Indiana for this purpose in 1989. An additional 2,000 tons was used for sand blasting bridges in 1989.

Fly Ash

This third waste product from coal fired plants, fly ash, is the particulate ash material that is light enough to be carried up the stacks. Fly ash has been used as a partial replacement for Portland cement in concrete construction. Roughly 10,000 tons of fly ash were used in concrete pavements in 1989.

Crumb Rubber

INDOT is currently experimenting with the use of ground tire rubber in bituminous construction and as a crack sealer. One contract on I-465 has been implemented to evaluate the use of a wet process (see Section 3.1.2 of this report), where rubber is added to and reacts with hot asphalt cement. Another project is being planned to evaluate the dry process (see section 3.1.2). An ongoing study is monitoring the performance of asphalt-rubber as a crack sealer for bituminous pavements.

Other Materials

Several other waste materials are allowed by the Standard Specifications, but are rarely chosen for use by the contractors. These include the following: reclaimed concrete for use in concrete construction or for rip rap (slope protection); and steel slag for compacted aggregate shoulders, rip rap, aggregate for bituminous construction, and snow and ice abrasives.

2.1.4 Research into Future Uses

INDOT has been actively researching potential uses of waste materials in highway construction and operations. In response to the environmental laws that mandate the use of waste materials in the highway industry, INDOT has initiated a number of research studies searching for practically sound, economically beneficial, and environmentally acceptable uses of waste materials in highway construction. The current study is another step towards the achievement of above stated goals. The other ongoing research projects are described below.

<u>The Role of INDOT in the Control of Hazardous Materials</u>

This study, being conducted by Professor J. D. Fricker and Professor Corson at Purdue University, West Lafayette, Indiana, will provide technical assistance to aid compliance of INDOT with employee and environmental protection regulations (McDaniel, 1990). The Pollution Prevention portion of the project will help INDOT to identify the waste stream and make recommendations on how to reduce it. It will recommend how source reduction can be accomplished through substitution for currently used materials, promote better "housekeeping" techniques and encourage on site recycling. This study will be completed by December 31, 1991.

<u>Asphalt Additive to Control Rutting and Cracking</u>

This study is being conducted by Ms. R. S. McDaniel, Division of Research, INDOT, West Lafayette, Indiana to assess the performance of bituminous mixtures modified by the addition of polymer additives (McDaniel, 1990). Asphalt-rubber (wet process; see Section 3.1.2) is one of the seven additives under study. Another additive, polyethylene, could use waste milk jugs if a consistent, reliable source of reclaimed polyethylene could be established. At present, the polyethylene used is scrap from plastic manufacturers. The test sections were placed on I-465 in September 1990. Phase I of this project is scheduled for completion in February 1993. Phase II, if approved, will continue the evaluation until failure of the pavement sections.

Evaluation of Crack Sealant Performance on Indiana's Bituminous Pavements

This study is being conducted by Mr. David Ward, Division of Research, INDOT, West Lafayette, Indiana, to evaluate the performance of several different types of sealants for cracks in bituminous pavements (McDaniel, 1990). Rubberized asphalt is included in this study.

Evaluation of Coal Bottom Ash

The INDOT sponsored a study to determine the feasibility of using bottom ash in highway construction. The study was conducted by two graduate students, Mr. T. -C. Ke and Dr. W. H. Huang, under the directions of Professor C. W. Lovell at the Purdue University (Huang, 1990; Huang and Lovell, 1990; Ke, 1990; and Ke et al., 1990). The laboratory study was performed in three phases: ash characterization tests, engineering properties tests, and environmental evaluation. The results of this study are reported in Section 3.4.3 of this report.

2.1.5 Conclusions

INDOT has been using recycled or waste products for many years in those applications which have been proven effective. They have selected those materials which have resulted in performance equal to, or superior to, that of conventional virgin materials. Their research program demonstrates their commitment for identifying additional materials that can be used without detriment in future construction, maintenance and operational applications.

2.2 United States Highway Agencies

2.2.1 Introduction

To obtain information on the current practices in the United States in the use of waste materials in highway construction, a questionnaire was developed and distributed to all the state highway agencies. The questionnaire was designed to seek information on the following three aspects:
- the types of waste materials currently used in highway construction, their applications, annual quantities, the state experiences in their uses for respective

applications from technical, economic, and environmental viewpoints, and the materials and applications which appear favorable and are to be projected for future construction;
- the materials and uses mandated by state laws - to determine whether the material being used is an attractive economic alternative to conventional materials;
- the types of waste materials that are used in a process covered by patents - to determine the possibility of further research and development.

Of the 52 questionnaires distributed, 42 were returned indicating a return ratio of 80.8%. Most of the states have expressed an interest in learning the results of this study. Besides providing answers to the specific questions, they have also sent their recent research updates and evaluation reports concerning the use of various waste materials in highway construction. A copy of the questionnaire is included in Appendix A. The results of survey questionnaire are summarized in Tables 2.1 - 2.6.

Table 2.1 presents the state-wise response to Questions 1-3, 5, and 6 of the survey questionnaire. Notes are included at the bottom of the table to explain the abbreviations, provide information on those waste products not included in the main table and additional remarks to clarify the tabulated information. The information contained in this table is then summarized in Tables 2.2 and 2.3 to present an overall view of the state-of-the-practice. Table 2.2 tabulates the waste products and their applications in highway construction showing the total number of respondents who have reported their use for the various applications. Table 2.3 presents the evaluation of various waste products from technical, economic, and environmental standpoints as reported by the respondents. The response to Question 4, which relates to the materials and their uses mandated by state laws, is summarized in Tables 4 and 5. Table 6 contains the waste products currently being used by the respondents in a process covered by patents.

2.2.2 An Overview of Current Practice

A total of 27 waste products have been reported by 42 state highway agencies who responded to the questionnaire. These are currently in use (and/or being studied experimentally) in a variety of highway applications. Of the 27 waste products, only 11 are presently used by more than 5% of the respondents, which include (in descending order of number of reported users): reclaimed paving materials, fly ash, rubber tires, blast furnace slag, steel slag, bottom ash, used motor oil, boiler slag, waste paper, mine tailings, and sewage sludge. The 6 waste products which are currently used (and/or being studied experimentally) by two of the total respondents are: building rubble, waste glass, sawdust, ceramic waste, incinerator residue, and highway hardware. The use of the remaining 10 waste products (see Note 2 to Table 2.3), currently being used (and/or are being investigated) by one of the respondent state highway agencies, are generally available in lesser quantities or their production is restricted to some specific geographical locations. The subsequent remarks pertain to only those 17 waste products which are tabulated in Table 2.3.

Current practice indicates that reclaimed paving materials, fly ash and rubber tires are used by a large number of respondents (i.e., 97.6%, 75.6%, and 69%, respectively). The use of blast furnace slag has also been reported by a significant number of respondent state highway agencies (35.7%). The use of steel slag, bottom ash, and boiler slag also seems fairly attractive for highway applications (reportedly being used by 17 to 20% of the respondents). The remaining above-mentioned products are less frequently used by the respondents.

The respondent state highway agencies have generally reported approximate annual quantities of waste materials currently used. The reported annual quantities of waste materials indicate that reclaimed paving materials, slags, and ashes are generally used in large quantities. The annual use of rubber tires, although being used (and/or being studied experimentally) by a large number of states, is generally in small quantities, with a few exceptions (Arizona, Oregon, and Vermont state highway agencies). This indicates that the use of tires in highway practice is generally in an experimental stage.

The evaluation of waste materials with respect to economic, performance, and environmental factors is generally reported as at least competitive with the conventional materials, satisfactory and acceptable, respectively, with some exceptions (see Table 2.3). The most varied experience is reported in the case of rubber tires. Of the 29 state highway agencies who reported the use of waste tires in highway construction, an average of 65% described their experience in the use of this product. In summary, 53% of those who reported their experience consider its use as uneconomical, 30% experienced poor performance, and 9.5% are doubtful about its environmental acceptability. The use of glass is reported as uneconomical by the only state highway agency which offered comments. In the case of steel slag, one state highway agency identified potential problems related to its expansive nature when used as an aggregate in portland cement concrete and also expressed doubts about its environmental acceptability. Some of the state highway agencies have also expressed doubts about the environmental acceptability of reclaimed paving materials, fly ash, blast furnace slag, and sewage sludge. The only state highway agency which reported experience in the use of incinerator residue considers it environmentally unacceptable.

The information contained in Table 2.4 indicates that the use of waste materials, in the majority of the respondent states, is **not** required by state laws. However, a number of state legislatures are presently considering required use of some waste products in highways to reduce waste disposal problems. This has stimulated research and investigations to determine the suitability of a number of waste products. This is reflected in the recent research studies and updates on research sent by a number of state highway agencies along with the completed survey questionnaire.

The waste products being used by some of the respondents in a process covered by patents are included in Table 2.6. A majority of respondents have reported the use of rubber tires in rubber modified hot mix asphalt (HMA) pavements (PlusRide^{TM} is the patented product in which large rubber particles are used as a substitute for a portion of the aggregate in a dry process; see Subsection 3.1.2) or in asphalt-rubber product (in which rubber is added to hot asphalt cement to produce asphalt-rubber binder that is used in a variety of asphalt-rubber products; see Subsection 3.1.2).

All the waste products, reported by the respondents to the survey questionnaire, were subjected to preliminary evaluation considering various factors including: the reported experience of state highway agencies, the quantities generated annually, and the past experience of the highway community in the use of waste products reported in the literature. As a result of preliminary investigation, 11 waste products were selected for further discussion, which include those waste products which are either generated in large quantities and whose use would have significant impact on the environment or those products which indicate significant potential, but information on which was provided by only a few respondents. The waste products described in some detail in Section 3, include rubber tires, waste glass, reclaimed paving materials, slags and ashes, building rubble, sewage sludge, and incinerated residue. The conclusions and recommendations, based on state-of-the-practice reported by the state highway agencies and discussion in the subsequent section, are summarized in Section 4.

State-of-the-Practice   13

Table 2.1: Current Uses of Waste Materials in Highway Construction (December 1990)

State	Reclaimed Paving Materials			Rubber Tires				Coal Fly Ash			Slags: B.Furn. (bfs); Bottom Ash (ba); Boiler (bs);Steel (ss); Foundry Waste(fw)			Waste Glass	Others/ Mineral Wastes
	Uses[1]	Qty[2]	Feasibility[3]	Uses[1]	Qty[2]	Fsblty.[3]	Uses[1]	Qty[2]	Feasibility[3]	Uses[1]	Qty[2]	Feasibility[3]	1,2&3	
(1)	(2)	(3)	(4)	(5)	(6)	(7)	(8)	(9)	(10)	(11)	(12)	(13)	(14)	(15)
Alaska	-	-	-	a	na	e/vg/g	-	-	-	-	-	-	-	-
Arizona	a,b,c,j(s)	100	ce/s/d-y	a,b	1	na/vg/d-y	j(cc)	30	ce/g/d-y	-	-	-	-	-
California	b,h	na	ce/s/ha-y	a,d	na	na-y	-	-	-	ss-f,h	na	na-y	-	4,5,9
Colorado	a,b	na	e/s/s	d	na	e/s/s	a,c	na	e/s/s	bfs-j(sb)	na	e/s/s	-	14/(a)
Connecticut	h,g,e,f,d ,a,c,b	na	ce/g/g	a	sq*	ue/g/d	-	-	-			Plane to use as 1c,b,d	-	14
Delaware	a,b,c	19.5	e/g/s-y	j(cs)	0.023	e/g/s-y	h	10cy	e/s/s-y	bfs-a (cc)	3	e/vg/s-y	-	-
District of Columbia	a,c,e	7	e/s/s	-	-	-y(a)	a,c,e	1	e/s/s	-	-	-	plan to use as1a	-
Georgia	f	na	ce/g/g-y	j(cs)	0.075	ce/g/g-y	j(p)	na	ce/g/g-y	bfs-e ba-g	na na	ce/g/g-y ce/g/g-y	-	-
Hawaii	b	1	na	-	-	-	-	-	-	-	-	-	-	-
Idaho	a,e	100	e/g/g-y	a	0.05	ue/g/g-y	a	0.1	ce/g/g-y	bfs-i,f,e	30	ce/g/en a -y(f)	-	5,7
Illinois	a,b,c,f,h	571.8	ce/s/d-y	j(cs)	na	ce/s/s- y	a,b,c	7.76	ce/s/s-y	bfs-e,f,g,i,j (cc) ba/bs-j(sc) ss-e,i	135.53 3.69 96.26	ce/vg/d-y ce/vg/s-y ce/vg/d-y	-	4
Indiana	rap-bj(s)	77	ce/g/g-y	a,b, j(cs)	sq*	e/na/s-y	a,d	10	ce/na/s-y	bfs-a,b ba-d bs-a,b ss-a	2000 sq 100 sq	ce/s/s-y na/s/s-y ue/s/s-y ue/s/s-y	-	4,5,9, 16
Iowa	rap-a,b,f rcp-a,b,f	200 250	ce/s/g ce/s/g	a	sq*	na	a,h	40	ce/s/g	-	-	-	-	-
Kansas	a,b,c,d,f	1100	ce/g/s-y	a*,b*, j(cs)	0.088	na/na/s- y*	a,b,c,d	24	ce/g/s-y	bs-a, j(ic,psa)	.04	na	-	-/21(c)

Table 2.1: Continued

(1)	(2)	(3)	(4)	(5)	(6)	(7)	(8)	(9)	(10)	(11)	(12)	(13)	(14)	(15)
Kentucky	a,b	>10	e/g/s-y	-	-	-	c,d, j(cc)	na	ce/g/g-y(b,d)	bfs-a ba-a,c bs-a, j (s,sc) ss-a	>40 sq* >20 >5	e/g/s-y na/g/s-n-y na-y na/g/s-y	-	12,15/ 21(b)
Louisiana	g,h	na	e/g/s	-	-	-	j(cc)	na	e/s/s-y				-	-
Maine	b,f	35	ce/s/s	j(sc)	*	na-y (a,b,h*)		-	-				-	-
Maryland	a,b	250	ce/s/s	-	-	-y(a)	j(cc)	.025	ce/s/s	bfs-j(cc)	30	e/vg/s	-	-
Massachusetts	b,c,f	60	ce/g/s-y	a	sq*	ue/p/s-n	j(cc)	na	ce/s/s-y	bfs-j(Newcem)	na	ce/g/s	-	-
Minnesota	a,b,c,f	large	ce/s/g-y	a,h (lwf)	0.03*	a-ue/p/s h-e/s/d	a(cc)	large	na-y	ss-a,b	na	na-y	-	5,11, 18*19*
Missouri	a	0-40	e/g/g-y	a,b	sq*	na-y	a,j(us)	13.89	e/g/g-y	bfs-e,f ba-a,bj (ic) bs-a,j (ic) ss-a,b fw-a,b	sq sq* 7.69 49.38 sq*	e/g/g-y e/g/g-n e/g/g-y e/g/g-y e/g/g-n	-	4,5,19
Montana	rap-b,c,d ,f,e rcp-j(pcc)	100m 7m	ce/s/d-y na-y	a	sq*	ce/vg/g-y	a,d	sq*	e/g/g	-	-	-	-	16
Nebraska	a,b,e,f	280	e/s/s-y	a	sq*	na/na/s-y	a,c,j	90	na/na/s-y	-	-	-	-	4
New Jersey	rap-a,b rcp-b,d	50 40	ce/g/d na/g/g-y(b,c)	j(cs)	sq	na	f	20	ce/g/g	bfs-g	sq*	na	b,na,na-y	6*,10, 16,18, 19,20
New Mexico	e,f	450	ce/vg/g-y	a,j(ar SAMI)	sq	a-na/p/g j-na/g/g	a	2	na/g/g-y	-	-	-	-	5
New York	b,d,h	na	ce/s/s	-	-	-	j(cc)	na	na/s/s-y(h)	-	-	-	-	-
North Dakota	rap-e,f,g rcp-e,f,g	150 225	na/g/g-y(a,g) ce/vg/g-y(f)	-	-	-	a	10	ce/g/g-y (a-cc)	-	-	-	-	-

Table 2.1: Continued

(1)	(2)	(3)	(4)	(5)	(6)	(7)	(8)	(9)	(10)	(11)	(12)	(13)	(14)	(15)
Ohio	rap-a,b,c rcp-b,g,f	large sq	e/s/s-y e/s/s-y	-	-	-	j(p)	sq	na/s/s-y	bfs-e,g,f ss-e,g,f, bs-g,f,j (jg)	bfs+ss- 2750; bs-sq	bfs-c/s/s-y ss-na/p/s-y na	-	-
Oklahoma	b,f	40	e/s/s-y	a,j(cs)	0.015	ue/p/s	a,b,d	50	ce/g/s-y	-	-	-	-	-/21(a)
Oregon	a,b,e,f,g	76	ce/g/g	a*,h	5.75*	na/g/g	a,b	4.28	ce/g/g	-	-	-	-	4,17
Pennsylvania	a,c	na	na	a,b, j(cs)	na	na-y	a,j(cc)	na	na	bfs:a,b,f,g,h ba:h,j(as) bs:h,j(as) ss:a,c,f,h	na	na	22	6,7,16, 18
Rhode Island	rap-b rcp-c	14 10	na-y na-y	a	sq*	ue/s/ha	-	-	-	-	-	-	-	-
South Carolina	j(0m)	sq*	na-y	-	-	-	-	-	-	-	-	-	-	-
South Dakota	f	150	ce/g/g-y	-	-	-	a	10	ce/vg/g-y	-	-	-	-	-/21(a)
Tennessee	b,c,d,f,g, h	na	e/s/g	-	-	-	-	-	-	-	-	-	-	5,13,16
Texas	a,b,c,f	na	na	a,b,c,f	na	na	a,b,c,f	na	na	ba-a,b,c,f	na	na	-	8
Utah	a,g	na	na/ha/d-y	-	-	-,y(h)	a	na	na	-	-	-	-	-/21(a)
Vermont	a,b,c,f,g	8cy	na	h	3cy	ue/p/s- y(sf)	-	-	-	-	-	-	-	-
Virginia	a,b	300	e/s/s-y	a	.018*	ue/s/s-y	-	-	-	-	-	-	a,b,sq*, ue/na/s	-
Washington	a	10	ce/s/g-y(rcp)	a	<0.1	ue/p/g	j(cc)	1-5	e/g/g-y	-	-	-	-	-
West Virginia	a,b	na	na-y	-	-	-	a,j(cc)	7.9	na-y	bfs-a,f,h, j(pcc) ba-j(ic) bs-j(ic) ss-j(ic)	na na na 1.7	na-y	-	7
Wyoming	a,b,f	200	s/vg/g	j(cs)	.002	na/s/ha	-	-	-	-	-	-	-	-

Notes:
1. The material is used as: (a) Additive to wearing course. (b) Additive to base course. (c) Additive to subbase course. (d) Additive to subgrade/embankment course.

Table 2.1: Continued

(e) As a wearing course. (f) As a base course. (g) As a subbase course. (h) As a subgrade/embankment course.

(i) For landscaping. (j) Others (abbreviations used stand for: sc-seal coat, cs-crack/joint sealer, sb-sand blasting, p-pozzolan, cc- cement concrete, ic-ice control, psa-patch segregated areas, s-shoulders, un-under seal, rap-reclaimed asphalt pavement, rcp-reclaimed concrete pavement, ar-asphalt rubber, SAMI-stress absorbing membrane interlayer, bm-binder material, sf-slope flattening, as - anti-skid, lwf-light weight fill)

2. The quantity of material used annually: (...x1000 tons)/(...x1000 cy)-the quantity in tons/cubic yards; (sq) -small quantity; (na) - material has been used but quantity not available; (...*)- the material used for testing purpose only; abbreviations used, m-mile,

3. State experience in the use of waste material:

    ce,e,ue,na /   cost effective/equal/uneconomical/information not available;

    vg,g,s,p,na /   performance has been very good/good/satisfactory/not satisfactory/information not available;

    g,s,ena,d,na -   the use of material from environmental viewpoint is: good/satisfactory/environmentally not acceptable/environmental acceptability is doubtful/information not available.

    y,n (...)   the use of material is rated as favorable and is projected for future construction/not favorable (future uses of the material, if other than the current practice).

4. Waste Paper: (a) California , recycle; (b) Illinois,under study for landscaping, has no adverse environmental effects, is uneconomical; (c) Indiana, recycle; (d) Missouri, recycled paper used as a mulch overspray, is economical and environmentally feasible; (e) Nebraska, for landscaping; (f) Oregon, for landscaping as mulch.

5. Used Motor Oil: (a) California, recycle; (b) Idaho, used as additive to subbase course; (c) Indiana, recycle; (d) Missouri, recycle; (e) New Mexico, as asphalt plant burner fuel; (f) Minnesota, as fuel for bituminous plant; (g) Tennessee, fuel for asphalt plant, is economical, and environmentally safe.

6. Ceramic Waste: (a) New Jersey, being used as additive to base course in small quantity for testing purpose only; (b) Pennsylvania, as additive to wearing course and subbase course, as subbase (if graded property), and for pipe bedding.

7. Sawdust: (a) Idaho, 150,000 c.y. used as a subgrade/embankment course, also intended to be used in future for stabilizing embankment slides; (b) Pennsylvania, plans to use wood chips for compost; (c) Washington, as subgrade/embankment course.

8. Phosogypsum: Texas.

9. Highway Hardware: (a) California-guard railing, bridge signs, light standards, signals etc. are stored and reused on new construction or for maintenance; (b) Indiana, traffic signs, delineators are recycled.

10. Recycled Steel in Rebar: being used by New Jersey.

11. Ground Shingle Manufacturing Scrap: used 10 tons annually by Minnesota as additive to wearing course for testing purpose only.

12. Scrubber Sludge: used by Kentucky as subgrade/embankment material for experimental purpose.

13. Phosphate Slag: Tennessee, as aggregate in wearing course, performs well, is cost effective, and environmentally acceptable.

14. Building Rubble: (a) Colorado, as a subgrade/embankment course; (b) Connecticut, as additive to base and subbase courses, is cost effective and environmentally feasible; (c) New Jersey, plans to use as additive to base course.

15. Atmospheric Fluidized Bed Combustion (AFBC): used by Kentucky as additive to subbase and subgrade/embankment for experimental purpose.

Table 2.1: Continued

16. Plastic Waste: (a) Indiana, recycle; (b) Montana, ground polyethylene plastics used as additive to asphalt paving material on experimental basis, the process is patented as "NOVOPHALT" (c) New Jersey, intends to use recycled plastic as substitute to wood; (d) Pennsylvania, plans to use as plastic polysheeting, plastic construction fence, plastic liners/under-liners; (e) Tennessee, plans to use for fence post and delineator post

17. Straw: Oregon, for landscaping.

18. Sewage Sludge: (a) Minnesota, as additive to wearing course and base course, 50 tons used for testing purpose, gives satisfactory performance at competitive cost, environmental acceptability is doubtful; (b) New Jersey, as compost, is economical, performance not evaluated, environmental acceptability is doubtful; (c) Pennsylvania, for landscaping as fertilizer and soil aeration.

19. Incinerator Residue: (a) Minnesota, tested as additive to wearing course, its environmental acceptability is doubtful; (b) Missouri, experimenting its use as additive to wearing course and base course; (c) New Jersey, intends to use as additive to base and subbase courses.

20. Shredded Wood Mulch: used by New Jersey for landscaping.

21. Mineral Wastes:
    (a) Mine Tailings: (i) Colorado, as subgrade/embankment course, performance, economics and environmental acceptability satisfactory; (ii) Montana, limited quantity used as a fill material; (iii) Oklahoma, uses lead/zinc mine spoils in polymer fast-set emulsion slurry; (iv) South Dakota, uses 75000 ton gold mine tailings annually as subgrade/embankment material; (v) Utah, as fill material.
    (b) Lime Kiln Dust: Kentucky, used as additive to subbase and subgrade/embankment, is environmentally acceptable, cost effective, and gives good performance, the above uses are favorable and projected for future construction.Its use in subbase is patented (N-Virocrete).
    (c) Kansas State uses 105,000 tons Chat, which is waste from lead and zinc mines that used to operate in SE Kansas, SW Missouri, and NE Oklahoma, as additive to wearing course and base course, gives good performance in SE Kansas, is cost effective and environmentally acceptable.

22. Pennsylvania reports that following materials and uses are favorable and are projected for future construction:
    Glass: as glascrete, glasphalt, aggregate, and skid resistant glass overlays
    Compost - used as mulch hydromulch

23. The information contained in this table is provided by respective state highway agencies in response to the survey questionnaire of this study (see Appendix A).

18 Use of Waste Materials in Highway Construction

Table 2.2: Summary of Waste Materials and Their Current Uses in the United States Highway Industry

Waste Material	No. of States Using the Material[1]	Material is Used as Additive to[2]:				Material is Used as[2]:				Landscaping (see 2)	Others (see 2,3)
		Wearing Course	Base	Subbase	Subgrade/ Embankment	Wearing Course	Base	Subbase	Subgrade/ Embankment		
Reclaimed Paving Materials	41	23	26	14	5	8	16	8	5	-	3(sh)
Coal Fly Ash	31	20	5	6	4	1	2	-	2	-	9 (cc), 1(us)
Rubber Tires	29	21	6	1	2	-	1	-	3	-	11 (cs)
Blast Furnace Slag	15	4	2	-	-	3	5	3	2	2	4 (cc), 1(sb)
Steel Slag	9	4	2	1	-	1	2	-	2	1	1 (ic)
Coal Bottom Ash	7	2	2	-	1	-	1	1	1	-	3 (ic), 1 (sc)
Used Motor Oil	7	-	-	-	1	-	-	-	-	-	3 (recy), 3 (apf)
Boiler Slag	7	4	1	-	-	-	1	1	1	-	3 (ic), 1(sc)
Waste Paper	6	-	-	-	-	-	-	-	-	5	1 (recy)
Mine Tailings	5	-	-	-	-	-	-	-	5	-	-
Sewage Sludge	3	1	-	-	-	-	-	-	-	1 (sa),2 (c/f)	-
Building Rubble	2	-	1	1	-	-	-	-	1	-	-
Waste Glass	2	1	2	-	-	-	-	-	-	-	-
Sawdust	2	-	-	-	-	-	-	-	2	-	-
Ceramic Waste	2	1	1	1	-	-	1	1	1	-	1 (pb)
Incinerator Residue	2	2	1	-	-	-	-	-	-	-	-
Highway Hardware	2	-	-	-	-	-	-	-	-	-	2 (recy)
Foundry Waste	1	1	-	-	-	-	-	-	-	-	-
Scrubber Sludge	1	-	-	-	-	-	-	-	1	-	-
Phosphate Slag	1	1	-	-	-	-	-	-	-	-	-
Straw	1	-	-	-	-	-	-	-	-	1	-
Plastic Waste	1	1	-	-	-	-	-	-	-	-	-
Lime Kiln Dust	1	-	-	-	1	-	-	-	-	-	-

Notes:
1. Of the 42 states who responded to the questionnaire.
2. The number under each column shows the total number of states that currently use the material in the respective application.
3. Abbreviations used: sh-shoulders, cs-crack sealer, cc-plain/structural cement concrete, us-under seal, ic-ice control, sc-seal coat, sb-sand blasting, recy-recycling, apf-asphalt plant fuel, pb-pipe bedding, f-fertilizer, c-compost, sa-soil aeration.
4. Information summarized in this Table can be found in Table 2.1.

Table 2.3: Evaluation of Waste Products from Technical, Economic, and Environmental Factors

Waste Materials	Total States[1]	Economic				Performance					Environmental				
		Evaluation by	Cost Eff.	Equal	Uneconomical	Evaluation by	Very Good	Good	Satisfactory	Poor	Evaluation by	Good	Satisfactory	Not Acceptable	Doubtful
Reclaimed Paving Materials	41	31	19	12	-	34	2	15	17	-	33	15	14	-	4
Fly Ash	31	22	15	7	-	24	1	13	10	-	26	12	13	-	1
Rubber Tires	29	15	3	4	8	19	3	5	5	6	21	8	11	-	2
Blast Furnace Slag	15	11	5	6	-	11	3	5	3	-	11	2	7	1	1
Steel Slag	9	3	1	1	1	5	1	2	1	1	5	1	3	-	1
Bottom Ash	7	3	2	1	-	5	1	3	1	-	4	2	2	-	-
Used Motor Oil	7	3, recycle; 3, use as fuel for asphalt plant (concern expressed about adverse effects on air quality); 1, additive to subbase.													
Boiler Slag	7	3	1	1	1	3	1	2	-	-	3	1	2	-	-
Waste Paper	6	2, recycle; 4, use for landscaping as mulch.													
Mine Tailings	5	-	-	-	-	-	-	-	-	-	-	-	-	-	-
Sewage Sludge	3	1	-	1	-	1	-	-	1	-	1	-	-	-	1
Building Rubble	2	1	1	-	-	1	-	-	-	-	1	-	1	-	-
Waste Glass	2	1	-	-	1	-	-	-	-	-	1	-	1	-	-
Sawdust	2	1	1	-	-	1	1	-	-	-	1	-	1	-	-
Ceramic Waste	2	-	-	-	-	-	-	-	-	-	-	-	-	-	-
Incinerated Residue	2	-	-	-	-	1	-	-	-	-	1	-	-	1	-
Highway Hardware	2	2, recycle.													

Notes:

1. Of the 42 states who responded to the survey questionnaire of this study.
2. The waste products whose use is reported by only one state highway agency are as follows: Foundry Waste, Phosogypsum, Recycled Steel in Rebar, Ground Shingle Manufacturing Scrap, Scrubber Sludge, Phosphate Slag, Atmospheric Fluidized Bed Combustion (AFBC), Plastic Waste, Straw, and Shredded Wood.
3. The information given in this Table is provided by the state highway agencies, based on their experience in the use of waste products, in response to the survey questionnaire of this study.
4. The information summarized in this Table can be found in Table 2.1.

Table 2.4: Materials and Their Uses Required by State Laws (December 1990)

State	Materials and Usage
California	Table 2.5 contains the recommendations of Caltrans for waste utilization in highways in response to the state law
Illinois	The use of coal fly ash is mandated by FHWA
Indiana	Not required, but must consider the use of waste materials.
Louisiana	Fly ash - additive to PCC as required by Federal law.
Maine	No, but state legislature is considering this issue.
Massachusetts	Fly ash and blast furnace slag are in the specifications by Federal decree, but not required to be used.
Missouri	(1) Rubber Tires: In bituminous Pavements on an experimental project only. (2) Used Motor Oil: In motor vehicles.
New Jersey	Fly ash - in LFA base course, FA concrete, Mineral filler to bituminous.
Pennsylvania	State laws mandate the PennDOT to conduct evaluation, testing and utilization of suitable waste materials.
Rhode Island	Required by Federal law to allow the optional partial replacement of Portland cement with fly ash for all field concrete except high early strength mixes, not used.

Note: This table contains the information provided by the state highway agencies in response to the survey questionnaire of this study (see Appendix A).

Table 2.5: Solid Waste Utilization in Highway Construction (after Caltrans 1990)

Solid Waste	Structural Section Material								
	AC	PCC	CTB	LCB	CTPB	ATPB	PM	AB	AS
Reclaimed Asphalt Pavement	OK	NO[1]	OK	OK	OK	OK	OK	OK	OK
Reclaimed PCC	OK[3]	OK[3]	OK	OK	OK	OK	OK	OK	OK
Foundry Slag	OK[2]	OK[3]	OK	OK	OK	OK	NO	OK	OK
Crumb Rubber	OK[3]	NO	NO	NO	NO	OK	OK	NO	NO
Ash	OK	OK	OK[3]	OK[3]	NO	NO	NO	NO	NO
Glass	OK[3]	NO	NO	NO	NO	OK	OK	OK	OK

Notes:

1. Unless data is presented indicating that all the specified PCC specification requirements will be satisfied.
2. Problems must be resolved regarding co-mingled blast furnace and steel slag.
3. With some limitations.
4. Abbreviations used: AC - asphalt concrete, PCC - portland cement concrete, CTB - cement treated base, LCB - lean concrete base, CTPB - cement treated permeable base, ATPB - asphalt treated permeable base, PM - untreated permeable material, AB - aggregate base, AS - aggregate subbase.

Table 2.6: Waste Products Used by State Highway Agencies in a Process Covered by Patents (December 1990)

State	Materials	Uses
Arizona	Asphalt Rubber	a,b
Colorado	Rubber Tires	Embankment
Kansas	Rubber Tires	a,b (asphalt-rubber, patented by ISI)
Kentucky	Kiln Dust(N-VIROCRETE)	Subbase
Massachusetts	Granulated Rubber Tires (pantented product of PlusRide™)	a
Minnesota	PlusRide™ (tested but do not use)	a
Missouri	Rubber Tires	a,b
Montana	Ground Polyethylene Plastics	a, (NOVAPHALT)
Nebraska	Rubber Tires	a
New Jersey	PlusRide™ (only tested in one application)	(not mentioned)
New Mexico	Rubber Tires	a, SAMI
Oklahoma	Mine Spoils	Slurry Seal
	Asphalt-Rubber	a, Joint Sealant
Oregon	Rubber Tires	a
Rhode Island	PlusRide™	a
Virginia	Rubber Tires	a
Washington	Rubber Tires, PlusRide™	a

Notes:
1. Abbreviations used in this Table stand for: a - additive to wearing course, b - additive to base course, SAMI - stress absorbing membrane interlayer.
2. Information contained in this Table is provided by respective state highway agencies in response to survey questionnaire of this study (see Appendix A).

3. Evaluation of Selected Waste Products

3.1 Rubber Tires

3.1.1 Background

The Rubber Manufacturers Association estimates that between 200 - 250 million worn-out car tires are generated each year (JAWMA, 1990). According to the Indian Department of Environmental Management (IDEM, 1991), over 11.5 million tons of waste tires are currently stockpiled in Indiana. The data summarized in Table 3.1 show that generation of rubber tires increased from 1.1 million tons in 1960 to 1.9 million tons (1.2% of total MSW) in 1988. Generation was higher in the 1970's and early 1980's, but the trend to smaller and longer-wearing tires has lowered the quantities. Projection show a modest growth in tonnage and nearly a "flat" percentage of total generation (see Table 3.1). Small amounts of rubber are recovered for recycling (5.6% was recovered in 1988). An estimated 240 million waste car and truck tires are discarded annually in the United States (Kandhal, 1990).

Tires occupy a large landfill space (due to low landfill density, i.e., 346 lb/cubic yard for rubber and leather as compared to 2268 lb/cubic yard for glass; EPA,1990). Disposal of large quantities of tires has accordingly many economic and environmental implications. Scrap tire piles which are growing each year pose two significant threats to the public:
- fire hazard - once set ablaze, it is almost impossible to extinguish (a tire store in Chicago burned for 6 weeks in May 1989 (Breuhaus, 1990));

Table 3.1: State of Tires Generation, Recovery, and Discards in MSW in United States (after EPA, 1990)

Year	Variations from 1960-88							Projected Estimates		
	1960	1965	1970	1975	1980	1985	1988	1995	2000	2010
Generated, Millions of Tons	1.1	1.4	1.9	2.5	2.6	1.9	1.9	2.0	2.1	2.2
Generated, % of Total Generation	1.3	1.3	1.6	2.0	1.7	1.2	1.2	1.0	1.0	0.9
Recovered, Millions of Tons	0.4	0.3	0.3	0.2	0.1	0.1	0.1	-	-	-
Recovered, % of Generation of Tires	36.4	21.8	15.8	8.0	3.8	5.3	5.6	-	-	-
Discarded, Millions of Tons	0.7	1.1	1.6	2.3	2.5	1.8	1.8	-	-	-
Discarded, % of Total Discards of MSW	0.9	1.1	1.4	1.9	1.9	1.2	1.3	-	-	-

- health hazard - the water held by the tires provides an ideal breeding ground for mosquitoes.

Methods for disposing of large quantities of scrap tires include their use as a fuel source (scrap tires in Minnesota are shredded and consumed as industrial boiler fuel (Mn/DOT, 1990 and Public Works, 1990)), as well as a raw material in production of other polymeric materials. Rubber Research Elastomers, Inc. of Minneapolis has developed a patented process for treating granulated tire rubber to produce a raw material for use in the production of other rubber products (Kandhal, 1990). The technology for the use of rubber tires in highway construction has been developed over the past three decades. Subsequent portions of this subsection will consider the feasibility of using waste tires, considering technical, economic, and environmental qualifications, for various applications in highway construction.

3.1.2 Use of Scrap Tires in Asphalt Pavements

"Crumb rubber additive" (CRA) is the generic term for the product from scrap tires used in asphalt products. It is the product from "ambient" grinding of waste tires and retread buffing waste. Tires can be ground by a "cryogenic" method, but the product is less suitable as CRA (Bernard, 1990). Addition of CRA to asphalt paving products can be divided into following basic processes:

- *Wet process* blends CRA with hot asphalt cement and allows the rubber and asphalt to fully react in mixing tanks to produce an asphalt-rubber binder. This binder can contain as much as 30% CRA. Both the wet process and the products which use the asphalt-rubber binder are protected by patents (Bernard 1990).

- ***Dry Process*** mixes CRA with the hot aggregate at the hot mix asphalt facility prior to adding the asphalt cement. This process produces a rubber modified hot mix asphalt (HMA) mixture. "PlusRide^{TM}" is the patented product of the dry process (Bernard 1990).

The four general categories of asphalt paving products which use CRA include: crack/joint sealants, surface/interlayer treatments, HMA mixtures with asphalt-rubber binder, and rubber modified HMA mixtures.

Crack/Joint Sealant

Crack/joint sealant may be an asphalt-rubber product, blending 15 to 30% CRA with the asphalt cement. It is covered in the American Society for Testing and Materials (ASTM) specifications (ASTM D3406). The results of the survey reported here show that 11 state highway agencies currently use asphalt-rubber as a crack/joint sealant. The performance of asphalt-rubber as a crack/joint sealant is reported to be satisfactory. Stephens (1989), based on nine-year evaluation of field performance of asphalt-rubber as joint sealant, reported that site-mixed materials performed better than pre-mixed materials, and that the winter sealing of concrete pavement joints was not as effective as summer sealing.

Surface/Interlayer Treatments

Surface/interlayer treatments may use an asphalt-rubber binder with 15 to 30% CRA. This application of CRA began in the late 1960's and was patented under the trade name SAM (Stress Absorbing Membrane) and SAMI (Stress Absorbing Membrane Interlayer).

- *SAM* is a trade name for a chip-seal with an asphalt-rubber sealant. The purpose of this layer is to seal the underlying cracks, thereby preventing the entry of surface water into the pavement structure. It is also intended to absorb the stresses that would lead the underlying cracks to reflect up to the surface. It is formed by applying asphalt-rubber on the road, covering it with aggregate and seating the aggregate with a roller. The thickness of the application usually varies from 3/8 to 5/8 in. (Singh and Athay, 1983), and 0.5 to 0.65 gallons per square yard of binder is applied to the surface. Another approach to the construction of a SAM is to proportion and mix the asphalt-rubber material and chips in a conventional asphalt hot mix plant and to place the resulting mixture on a grade with conventional asphaltic concrete spreading machine. However, the cast-in-place SAM's have performed better (Vallerga, 1980).

- *SAMI* is a layer, with an asphalt-rubber binder, sandwiched between the road base and an overlay. The only difference between SAM and SAMI is that SAM does not have an overlay whereas SAMI does. The intended purpose of SAMI is to reduce reflection cracking by cushioning or dissipating the stresses from the underlying pavement before they are transferred to the overlay. The procedure in placing the SAMI is similar to that used in placing the SAM, with a few differences in design aspects.

<u>Asphalt-Rubber Mixtures</u>

Since late 1960's, the use of asphalt-rubber binder in HMA mixtures has been researched. Two such process have been reported:

- McDonald Process - initiated in 1968, in which hot asphalt cement is mixed with 25% ground tire rubber to establish a reaction, and then is diluted with kerosene for easy application (Schnovmeier, 1986).

- Arm-R-Shield™ or Arizona Refinery Process - Initiated in 1975, and was patented by the Union Oil Company. It is currently marketed by Arizona Refinery Company (ARCO). The ARCO product incorporates extender oils and 18 to 20% recycled rubber from scrap tires directly in the hot liquid asphalt (Schnormeier, 1986). The reported benefits of using A-R-S modified hot mix surfacing include (Arm--R-Shield, 1986; reported by McQuillen and Hicks, 1987):
  (a) Flexibility down to -26° C (-15° F).
  (b) Higher viscosity than conventional asphalt at 60° C (140° F).
  (c) Tougher (in relation to surface wear from studded tires) and a more elastic surface.
  (d) Greater resistance to aging.
  (e) Recycling of used rubber tires.

Rubber Modified Asphalt Mixtures

The concept of introducing coarse rubber particles into asphaltic pavements (using the dry process) was developed in the late 1960's in Sweden. It was originally marketed by Swedish companies, Skega AB and AB Vaegfoerbacttringar (ABV) under the patented name "Rubit". This technology was introduced in the United States in the 1970's as the patented product, PlusRide™ and is marketed by All Seasons Surfacing Corporation of Bellevue, Washington (Bjorklund, 1979; Allen and Turgeon, 1990). The PlusRide ™ process typically uses 3% by weight granulated coarse and fine rubber particles to replace some of the mix aggregates (Bjorklund, 1979). The reported advantages of using the PlusRide™ in HMA applications are (PlusRide™ 1984; reported by McQuillen and Hicks, 1987):

- Reflective and thermal pavement cracking are greatly reduced.

- Resistance to studed tire wear is increased.
- Skid resistance is increased.
- Ice removal by deformation of the rubber granules under traffic loading and vehicle generated wind.
- Suppression of pavement tire noise.
- Recycling used rubber tires which are currently a major environmental problem.

3.1.3 Experience in the Use of Rubber Modified Asphalt Paving Products

<u>Alaska</u>

The Alaska DOT has conducted extensive laboratory and field studies on the use of rubber modified asphalt. The results of these evaluations have been published in the form of reports and papers (e.g., Esch, 1984; Takallou et al., 1985; Takallou et al., 1986; McQuillen et al., 1988; Takallou and Hicks, 1988; Takallou et al., 1989). Salient conclusion from some of these studies are summarized below.

Esch (1984) reports the evaluation of six experimental rubber-modified pavement sections, totaling 3.4 miles in length, constructed between 1979 and 1983. In these projects, 3 to 4% of coarse rubber particles were incorporated into HMA using the PlusRide™ process (dry process). Salient findings and conclusions of this study are as follows.
- The attainment of an average field voids level of less than five percent, with maximum voids below 8% are critical to pavement resistance to raveling.
- Field voids of less than 5% are highly desirable.
- The benefits of rubber-modified paving mixes include: the ability to shed an ice cover more quickly than conventional pavements, the development of a more

flexible and fatigue resistance pavement, a reduction in tire noise, and recycling of used tires. Under Alaskan conditions of icy non-salted roadways, stopping distances were consistently reduced by the use of rubber modified asphalt pavements, averaging 25% less than on normal pavements.

Another study, conducted by Takallou et al. (1985), reports the results of a research project which includes a survey of field performance and laboratory evaluation of mix properties as a function of a number of variables, such as rubber gradation and content, void content, aggregate gradation, mix process, temperature, and asphalt content. They evaluated twenty different mix combinations at two different temperatures (-6° C, + 10° C). The results of this study have been used to develop mix design recommendations for rubber modified asphalt mix for use in Alaska (interested readers can use these mix design recommendations as guidelines for further study and development of mix design suitable for local environments). Their 1984 field survey results indicated that most rubber modified pavements placed to date have not failed in fatigue. Where problems had been reported, they had generally been early raveling, and were attributed to excessive voids resulting from poor compaction and/or low asphalt content.

McQuillen et al. (1988) presented economic analyses showing that the use of rubber modified asphalt products is cost effective compared to the conventional HMA in Alaska, based on life-cycle costs.

Arizona

The use of asphalt-rubber by the Arizona DOT has been reported by Morris and McDonald (1976) and Schnormeier (1986). Since January 1967, the Arizona DOT has used asphalt-rubber in a variety of ways for pavement seal coats, SAM, SAMI, subgrade seals, lake liners, joint and crack fillers, roofing, and airport runway surfacing. Morris and McDonald (1976) based on the performance of surface treatments (two SAM and a SAMI) using asphalt-rubber mixtures on three pavement sections and a laboratory study, reached two conclusions: (1) asphalt rubber products, when placed as SAM, controlled reflection of fatigue cracks and was an effective alternative to a major overlay or reconstruction; (2) when placed as SAMI, the system effectively controlled reflection of all cracks.

Schnormeier (1986) evaluated the performance of asphalt-rubber placed as SAM, SAMI, and crack sealant between 1969 and 1974 in Pheonix, Arizoña. It was concluded that asphalt-rubber stops reflective cracking in paving materials with cracks less than 0.25 in. in thickness for 8 to 12 years. It also waterproofs the surface to obtain maximum stability; seals the subgrade to minimize volume changes, and is an excellent crack filling material and joint sealer. The cost analysis showed that asphalt-rubber placed as SAM costs twice as much as conventional chip-seal. However, the study also concluded that 10 - 12 years of maintenance free life can be expected from an asphalt-rubber seal, whereas conventional chip-seal can last for 6 - 8 years, with some maintenance. Hence, the life cycle costs of the two products will be about equal.

Connecticut

ConnDOT's (Connecticut Department of Transportation) report on "Eight-Year Evaluation of an Asphalt-Rubber Hot Mix Pavement" (Larsen, 1989a), describes the performance of an experimental 900 ft section of asphalt-rubber hot mix bituminous pavement laid on State Route 79, in Madison, in October 1980. The asphalt-rubber binder consisted of 20% finely ground rubber from the Arizona Refining Company and conventional asphalt (grade AC-20). The pavement was placed as a 1.5 in. thick overlay in one lift using conventional paving technique. A standard ConnDOT class 2 bituminous pavement was placed at the same time and used as a control section.

An evaluation conducted during summer of 1989 concluded that in comparing the asphalt rubber pavement to the control section:
- it performed better with respect to transverse, longitudinal and alligator cracking;
- it showed slightly lower skid resistance, but the friction level remained adequate.

Another report "Nine-Year Evaluations of Recycled Rubber in Roads" (Stephens, 1989) described the use of asphalt-rubber in Connecticut in various forms of pavement rehabilitation, including: thick overlays (4 cm), thin overlays (1.5 cm), chip seals, crack and joint sealing, and stress relieving interlayers. Their experience with the use of asphalt-rubber in various products is briefly given below.

- *Thick Overlays* - the use of asphalt-rubber on thick overlays in various percentages of the total mix indicated that one percent rubber reduced the longitudinal cracking of overlays placed over medium to slightly distressed pavements. In contrast, the amount of cracking over highly distressed pavements, regardless of traffic level, was greater in the asphalt-rubber section. Due to the random performance of test sections, they could not conclude whether asphalt-rubber sections performed better than the control. Their experience with 2% CRA in overlays (with the exception of one section) indicated that the asphalt-rubber overlay developed cracking twice as fast as the control sections.
- *Chip Seals* - for 3- to 4-year evaluations, asphalt-rubber sections performed better than the control sections. At nine-year evaluation, all sections had been covered. The new surface placed over asphalt-rubber sections show less cracking.

- SAMI's reportedly did not show improved performance of any section.

Finally, the report concludes that the asphalt-rubber products mentioned above did not prove greatly effective, with the exception of seal coats.

Kansas

The Kansas DOT first utilized rubber-asphalt subbase mixture in SAMI and crack sealing in mid-1970's (McReynolds, 1990). The SAMI's were placed over pavements which had severe transverse cracking. The use of SAMI did not significantly delay or reduce the reflection of cracking, hence, its use was discontinued. However, the use of asphalt-rubber as a crack sealant proved successful and has become the standard crack filling material used in Kansas on moderate width cracks. The Kansas DOT has planned the use of asphalt-rubber on two projects in which overlays of varying thickness will be laid on distressed pavements.

A recent study by Kansas DOT on the "Economics of Using Asphalt Rubber in Pavement" (KDOT, 1990) show that the use of this product in HMA is highly uneconomical. Their economic analysis, based on current prices of various products in Kansas and the quantity of crumb rubber used per ton of HMA (i.e. a typical ton of asphalt-rubber hot mix contains approximately 25 lbs ≈ 1.5 rejected tire) shows that the disposal of each tire by this method would cost $11 to the state.

Maine

The Maine DOT (1990) reports the use of rubber in asphalt mixtures on the following four project since mid 1970's.

- In 1976, a stress relieving, rubberized slurry interlayer was placed on ten lane-miles of I-95 in Carmel. A cold mixture of shredded rubber tires, sand, and emulsified asphalt was used. Cracks appeared in the test section after the first winter.
- In a second case, they attempted to use asphalt-rubber in an overlay in West Gardiner, the construction of this overlay could not be completed due to serious construction difficulties as a result of excessive material viscosity.
- In 1988, the Maine DOT applied asphalt-rubber chip seals to cracked pavement in Peru and Kennebunk. The performance of test sections was superior to control sections with respect to reflection of cracks, but the test section laid in Peru showed excessive raveling and had to be conventionally overlaid after about one year.
- The fourth project was a rubber chip seal constructed by the Federal Aviation Administration at the Norridgewock Airport in 1979. In view of the excellent performance of the test section, the entire runway was chip-sealed using an asphalt-rubber mixture in 1982. No significant distress was observed after four years. However, the condition of the original pavement is not a matter of record.

Minnesota

Mn/DOT (1990) shows that the department has researched the use of asphalt - rubber in seal-coat, interlayer, crack sealing, and asphalt concrete systems, as well as the use of rubber modified asphalt concrete. They have recently constructed a 2-mile test section with asphalt paving mixtures which contained varying percentages of recycled tire rubber and shingle scrap (Mn/DOT, 1990a). Their preliminary report states that the use of shingle scrap was identified as a means of reducing the asphalt demand of the mixtures and make the use of rubber more economical (performance and economic analyses of this experimental project are reportedly under way).

Mn/DOT's report on the use of asphalt-rubber products in Minnesota (Turgeon, 1989), evaluates the performance of asphalt-rubber products. Based on the evaluations of two SAM projects, three SAMI projects, and one project employing asphalt-rubber as a binder in a dense graded mix, the report concluded that:

- SAM's may prove cost competitive for certain condition, if laid properly;
- SAMI's do not eliminate but do decrease the reflection of cracks; however the benefits do not appear to justify the additional cost;
- the asphalt-rubber and the conventional overlay experienced an equal amount of cracking, and the 100% increase in price over conventional mixtures does not appear to be justified.

Mn/DOT recently issued a report on the "Evaluation of PlusRide ™ (A Rubber Modified Plant Mixed Bituminous Surface Mixture)" (Allen and Turgeon 1990). They constructed two experimental projects in September 1984 using a rubber modified asphalt mix. Both test projects were four-lane divided highways and had a two-way average daily traffic of about 10,000, with 17% being truck traffic. The findings of this study, which were based on evaluations of design, construction procedures/behavior of mixes during construction, and post-construction performances, include:

- design procedures are not fully developed;
- mixtures are more susceptible to compaction when adverse weather or equipment problems occur;
- compared to the control mix, rubber modified mixtures have slightly lower friction numbers, slightly rougher surfaces based on ride meter (Mays) measurements, almost equal surface deflections when tested with the Falling

Weight Deflectometer (FWD), equal tire noise levels, and showed no significant de-icing benefits.

One test section raveled severely and was removed and replaced in 1985. The second test section, although its surface is somewhat more ragged than the control mix, has performance equal to the control mix. The study concluded that the rubber modified material, which cost twice that of the conventional mixture, had not displayed any significant benefit to the pavement, and recommended that further use of this material should be discontinued.

New York

The New York State Department of Transportation (NYSDOT) specifications allow the use of asphalt-rubber materials for liquid joint and crack sealing (NYSDOT, 1990). NYSDOT completed two resurfacing test projects in the summer of 1989. Test sections were placed on Route 144 in the Town of Bethlehem, Albany County and on Route 17 in the Town of Deposit, Delware county at a cost of 50% and 114% more than the conventional asphalt mixes, respectively. At both sites, each of five different mixes were applied in separate adjoining 2,000 ft sections of highway as follows: 1, 2, 3% crumb rubber from New York state waste tires, respectively, 3% CRA from PlusRide™, and a conventional asphalt concrete mix, which was used as the experimental control (NYSDOT, 1990).

The above report states that the factors which cause rubber modified asphalt to be more costly than conventional asphalt mixes include: cost of the granulated rubber, the need for a more costly aggregate (stone) and filler gradation, increased energy to heat the asphalt mix to the higher temperature required for a rubber modified mix and to extend the mixing time to assure proper mixing, increased plant labor to handle the rubber additive and increased labor and equipment costs at the highway work site. Their experience with the test projects shows that the rubber mixes tend to be "sticky", adhering to the equipment and making release from the delivery truck bed more difficult, and requiring extra care with water and additives in rolling to prevent adherence to the rollers.

The stickiness reportedly increases with increased rubber content. Difficulties have also been reported in obtaining a rubber-modified asphalt with an acceptable gradation, since it is difficult to procure suitably "gap graded" conventional aggregates to accommodated the rubber particles. The report also expresses concerns about increased air pollution as a result of adding rubber to the mix and also the higher temperature required during mixing. The post-construction performance evaluations of these projects is under way.

Pennsylvania

The Pennsylvania DOT (PennDOT) has reported results of two research projects which compared the performance of SAMI's placed using asphalt-rubber binder (33% CRA) with control sections containing conventional asphalt binder (Mellott, 1989). Both projects involved base repair, a leveling course, SAMI, and 0.5 in. of ID-2 wearing surface course material. However, in one project an additional 2 in. layer of ID-2 binder course material was placed between SAMI and ID-2 wearing course material. The control sections were randomly placed. However, except for one section, SAMI was not placed on the control sections. Hence, the comparison was mostly based on pavements with SAMI laid in asphalt-rubber binder and without SAMI.

It is reported that the control section containing SAMI placed in a conventional asphalt binder had excellent performance after 8 years, whereas the section containing SAMI laid in asphalt-rubber binder failed after one winter. Based on the overall performance of all the test sections, the report concludes that the increased cost of the materials versus insignificant increase in the service life, does not economically justify the use of SAMI. The report also states that:

> "Pennsylvania has been evaluating the use of asphalt-rubber since the early 1960's. There have been numerous projects placed and evaluated without one major success."

Vermont

The Vermont Transportation Agency (1988) briefly describes their experience in the use of asphalt-rubber placed in surface treatments (SAM, SAMI). Test sections were placed on I-91 in the town of Springfield and Weathersfield. It was concluded, based on annualized cost and quality of performance, that application of asphalt-rubber for surface treatments was not cost effective.

Washington

A rubber modified asphalt experimental overlay was constructed in Washington (Mt. St. Helens project) in 1983. The project, constructed in the Gifford Pinchot National Forest, consisted of 1.11 miles of continuous test sections of three different thickness, i.e., 1.75, 2.5, and 3.5 in. An additional 3.5 in. thick overlay was constructed using asphalt concrete for comparison. Laboratory and field test results after three years show (Lundy et al. 1987):

- Moduli of asphalt rubber modified as well as conventional material are increased.

- Laboratory fatigue lives of both materials are decreasing with time. However, expected fatigue life of the rubber-modified mix exceed that of the control for any given strain level.
- Control mixture shows a greater increase in stability with time.
- Mays ride meter tests indicate the rubber modified section to be slightly rougher.
- When tested dry, the control section has higher skid numbers.
- Indirect tensile tests indicate the control mixture has greater strength.

In response to the survey questionnaire (Appendix A), Anderson (1990) offered the following comments on the Washington DOT's experience in the use of rubber modified asphalt products:

"...Our state has not had a good experience with the proprietary product PlusRide. About 50% of the projects constructed with this product have experienced premature failure or severely shortened service life. In other more successful installations the product has not shown the superior performance promised by the product literature and the claims of suppliers. We currently have a moratorium on the use of this product on state owned highways."

3.1.4 Discussion

Various laboratory and analytical studies (Kekwick, 1986; Lundy et.al, 1987; McQuillen et.al, 1988; Takallou et.al, 1985; 1986; 1988 and 1989; Vallerga, 1980) and industry publications (e.g. PlusRide™ 1984; Arm-R-Shield, 1986) show that adding CRA to asphalt paving products (as a binder or as an aggregate) improves the engineering characteristics of the pavements, including the service life. However, these claims are not always substantiated by the field performance of asphalt paving products containing CRA.

The experience in the use of CRA in asphalt paving products, as described in the preceding subsections, showed both successes and failures.

The intended purpose of describing the experience of a number of states in the use of CRA in asphalt paving products was to establish the basic causes of observed failures. However, it appears that with a few exceptions, the failures and successes have been random and no definite reasons can be offered with confidence for this unusual behavior (i.e. same percentage of CRA used in a similar product, under similar climatic environments demonstrated different behavior - one fails within a short period of construction, whereas the other performs much better than the control sections). Various reasons have been offered for the inadequate performance of the products (e.g., NYSDOT, 1990; ODOT, 1990) The writer is of the opinion that more research (analytical, laboratory and field studies) is required to completely understand this technology.

The asphalt paving products with CRA have also demonstrated consistently better performance in some states e.g. Alaska (rubber modified asphalt) and Arizona (asphalt-rubber). Similarly, some of the asphalt paving products have displayed better performance in most of the cases and suffered fewer failures, which include two products that use asphalt rubber binder, i.e., joint/crack sealant and SAM's.

Various studies on the economics of using CRA in asphalt paving products (e.g. KDOT, 1990; McQuillen et al., 1988; NYSDOT, 1990) show that the products are not cost effective, since the performance of the products is generally not commensurate with enormous increase in cost (the increase in cost is generally 50% to more than 100% higher than the conventional materials). However, the additional cost of asphalt-rubber binder as a joint/crack sealant is justified in view of better performance. Similarly, additional costs of materials used in SAM's has also been acceptable on the life cycle cost basis in most of the cases, due to its somewhat better performance and generally longer service life.

The asphalt paving products containing CRA are generally acceptable from environmental viewpoint. However, some concerns have been expressed over increased air pollution as a result of adding rubber to the mix and also the requirement of elevated temperatures during mixing.

The recycling of conventional asphalt pavements has gained wide popularity due to obvious economic and environmental benefits (see Section 2 of this report, and NCHRP, 1978; Ortgies and Shelquist, 1978; Wood et al, 1988 and 1989). Research studies have generally not addressed this issue (limited studies have been performed, but conclusions can not be generalized, e.g. Charles et al, 1980) in the cases of asphalt-rubber or rubber modified asphalt. If these pavements cannot be recycled on completion of their service lives, the disposal of these pavements will create another major waste disposal problem.

3.1.5 Use Of Tires in Subgrade/Embankment

Two techniques to incorporate waste tires in subgrade/embankment are: (1) use of shredded tires as a lightweight fill material; (2) use of whole tires or their sidewalls for soil reinforcement in embankment construction. Both of the techniques are practical and have been researched by some of the state highway agencies (see Table 2.1). Their experiences show that the use of tires in subgrade/embankment is quite promising, since significant engineering benefits are achieved, besides consuming large quantities of waste tires. The concept of using tires in subgrade/embankment is also extended to enhance the stability of steep slopes along the highways (TNR, 1985), temporary protection of slopes (Caltrans,

1988), retaining of forest roads (Keller, 1990), and protection of coastal roads from erosion (Kilpatrick, 1985).

Use of Tires as Lightweight Aggregates

Construction of roads across soft soil presents stability problems. To reduce the weight of the highway structure at such locations, wood-chips or sawdust have been used as a replacement for conventional materials. Wood is biodegradable and thus lacks durability. Conversely, reclaimed rubber tires are non-biodegradable and thus more durable.

The Oregon DOT used shredded tires as a lightweight fill and described the experience as a success (the department is currently preparing the report on this project; ODOT, 1990). The Mn/DOT has experimented the use of scrap tires in roadway fill across a swamp that is underlain with peat and muck. About 52,000 shredded tires were used as lightweight fill material in a 250-ft section of roadway. The section is reportedly performing satisfactorily (Mn/DOT, 1990 and Public Works, 1990c). Turgeon (1989) reports that employing waste tires as lightweight fill is a simple and cost competitive application which can use a significant amount of local tires.

Minnesota Pollution Control Agency (MPCA) sponsored a study on the feasibility of using "Waste Tires in Subgrade Road Beds" (MPCA, 1990). Twin City Testing Corporation (TCT), St.Paul, Minnesota, performed the laboratory study to evaluate the compounds which are produced by exposure of tires to different leachate environments. They subjected the samples of old tires, new tires and asphalt to laboratory leachate procedures at different conditions. They also conducted field sampling.

As a result of elaborate testing, TCT reached the following salient conclusions (MPCA, 1990):

- Metals are leached from tire materials in the highest concentrations under acid conditions, constituents of concern are barium, cadmium, chromium, lead, selenium, and zinc.
- Polynuclear Aromatic Hydrocarbons (PAHs) and Total Petroleum Hydrocarbons are leached from tire materials in the highest concentrations under basic conditions.
- Asphalt may leach higher concentrations of contaminants of concern than tire materials under some conditions.
- Drinking water Recommended Allowance Limits (RALs) may be exceeded under "worst-case" conditions for certain parameters.
- Co-disposal limits and EP Toxicity limits are generally not exceeded for the parameters of concern.
- Potential environmental impacts from the use of waste tires can be minimized by placement of tire materials only in the unsaturated zone of the subgrade.

MPCA (1990) states, based on a search of a number of databases, that they could not find any other published paper on leach tests on tire products or environmental assessments of the use of waste tires in embankments or subgrades.

The use of shredded tires in subgrade/embankment construction offer some technical and economic advantages under certain conditions. However, further research is required to evaluate the various factors, including environmental concerns.

Use of Tires for Soil Reinforcement

Various agencies have practiced and evaluated the use of tires for soil reinforcement. Forsyth and Egan (1976) described a method for use of waste tires in embankments and considered it a very promising application. The method involves separation of tire sidewalls and treads, the latter being a commercially valuable commodity. The tire sidewalls can be used as mats or strips in embankment to increase its stability. The laboratory and theoretical studies conducted by Caltrans (Forsyth and Egan, 1976) indicated that the systematic inclusion of tire sidewalls could possibly benefit a fill and thus permit steeper side slopes and increase resistance to earthquake loading.

Encouraged by the results of above mentioned study, Caltrans designed a tire-anchored wall system, in which tire side walls are used to anchor timber retaining structures (TRN, 1985, Caltrans, 1986). Designs are being developed to incorporate 6 ft. timber posts obtained from the removal and replacement of guardrail installations. This application is considered practical and very economical , but may have environmental implications as discussed above.

Turgeon (1989) describes the experience of the Minnesota Department of Natural Resources in the use of tires for soil reinforcement. They used whole tire mats and tire chunks as a material to replace corduroy logs in logging road embankments over swamps. This technology is reportedly spreading to other roadway projects.

The use of tires in retaining structures has also been practiced primarily for maintenance and rehabilitation of road embankments (Caltrans, 1988; Keller, 1990). Whole tires anchored in the backfill are used in various configurations for wall heights up to 10 ft. The application is economical, results in moderate face settlement and may have aesthetic and environmental implications. Another potential use of tires is in the form of reefs along the coast for prevention of scouring and protection of coastal roads.

3.2 Waste Glass

3.2.1 Background

The Environmental Protection Agency's report on "Characterization of Municipal Solid Waste in the United States: 1990 Update" (EPA, 1990) shows that waste glass constituted 6.7 million tons of Municipal Solid Waste (MSW) in 1960, or 7.6% of the total generation of MSW. Generation of glass continued to grow over the next two decades, but then glass containers were widely replaced by other materials, principally aluminium and plastics. Thus, the fraction of glass in the MSW declined in the 1980's, from 15 million tons in 1980 to 12.5 million tons in 1988. Glass was 10.0% of MSW generation in 1980, declining to 7% in 1988. The projected estimates demonstrate a continuous declining trend in the generation of waste glass. Conversely, an increase in the recovery of waste glass for recycling is predicted, from 1.5 million tons (i.e., 12% of total glass production) in 1988 to 2.1 to 3.1 million tons (i.e., 18.9 to 27.9% of total generation) in 1995. Waste glass is found in the MSW primarily in the form of containers (see Table 3.2).

Glass is composed mainly of silica or sand, but it also contains predetermined amounts of limestone and soda ash designed to produce uniform quality and color. There are three basic types of glass manufactured commercially in the United States: borosilicate, soda-lime, and lead glass. Approximately 90% of all glass produced is soda-lime glass (Miller and Collins, 1976). The chemical composition of the three basic types of glass is shown in Table 3.3.

Table 3.2: State of Glass Generation, Recovery, and Discards in MSW in United States (after EPA, 1990)

Year	Variations from 1960 to 1988							Projected Estimates			Projected Range, 1995	
	1960	1965	1970	1975	1980	1985	1988	1995	2000	2010	Low	High
Generated, Millions of Tons	6.7	8.7	12.7	13.5	15.0	13.2	12.5	11.1	10.3	9.5	-	-
Generated, %of Total Generation	7.6	8.4	10.4	10.5	10.0	8.2	7.0	5.6	4.8	3.8	-	-
Recovered, Millions of Tons	0.1	0.1	0.2	0.4	0.8	1.0	1.5	-	-	-	2.1	3.1
Recovered, % of Generation of Glass	1.5	1.1	1.6	3.0	5.3	7.6	12.0	-	-	-	18.9	27.9
Discarded, Millions of Tons	6.6	8.6	12.5	13.1	14.2	12.2	11.0	-	-	-	-	-
Discarded, % of Total Discards	8.1	8.9	11.0	11.1	10.5	8.4	7.1	-	-	-	-	-
Glass Packaging (Generation/Recovery/Discards, Millions of Tons)												
Beer and Soft Drink Bottles	1.4/0.1/ 1.3	2.6/0.1/ 2.5	5.6/0.1/ 5.5	6.3/0.4/ 5.9	6.7/0.8/ 5.9	5.7/1.0/ 4.7	5.4/1.1/ 4.3	3.8	2.8	2.0	-/1.3	-/1.7
Wine and Liquor Bottles	1.1/0.0/ 1.1	1.4/0.0/ 1.4	1.9/0.0/ 1.9	2.0/0.0/ 2.0	2.5/0.0/ 2.5	2.2/0.1/ 2.1	2.0/0.1/ 1.9	2.0	2.0	2.0	-/0.8	-/1.4
Food and Other Bottles and Jars	3.7/0.0/ 3.7	4.1/0.0/ 4.1	4.4/0.0/ 4.4	4.4/0.0/ 4.4	4.8/0.0/ 4.8	4.2/0.0/ 4.2	3.9/0.3/ 3.6	4.0	4.1	4.0	-/2.1	-/3.1
Total Glass Packing	6.2/0.1/ 6.1	8.1/0.1/ 8.0	11.9/0.2/ 11.7	12.7/0.4/ 12.3	14.0/0.8/ 13.2	12.1/1.1/ 11.0	11.4/1.5/ 3.6	9.8	8.9	8.0	-	-
Trend:	Year	1960-70	1970-80	1980-85	1886-88	88-2000	2000-10	-	-	-	-	-
Annual rate of inc. (or dec.) of generation of glass in MSW		7.0	1.7	-2.2	-1.9	-1.6	-0.8	-	-	-	-	-

Table 3.3: Chemical Composition of Glass (after Miller and Collins, 1976)

Constituent	Borosilicate	Soda-Lime	Lead
SiO_2	81	73	63
R_2O_3	2	1	1
Na_2O	4	17	7
K_2O	-	-	7
B_2O_3	13	Trace	-
CaO	-	5	-
MgO	-	3	-
PbO	-	-	22

The most obvious use for waste glass (commonly called cullet) is to recycle it to make new glass. This practice, besides reducing waste disposal problem, allows large savings in energy used for manufacturing of new glass from raw materials. Powell (1983) estimated that the use of 20%, 50%, and 100% cullet in glass manufacture would cause energy savings of 4%, 11%, and 22%, respectively. However, the entire waste glass generated cannot be re-used by glass manufacturers since only color-sorted and contamination-free cullet is considered feasible for re-use in the glass industry. Therefore, significant quantities of glass may be available for secondary applications.

Shortage of good quality materials, waste disposal problems, and shortage of conventional aggregates generated considerable interest in the past to examine the feasibility of the use of waste glass in highway construction. Various studies have indicated the potential for use of waste glass as aggregate replacement in portland cement concrete structures and pavements (e.g., ENR, 1972; Johnston, 1974; Breakspere, 1980). Asphalt pavements have also been studied (Byrum, 1971 and Watson, 1988). The glass-asphalt mixes in which glass replaces the conventional aggregates are called "glasphalt". Glass has also been used as unbound aggregate in base layers and as fill material in embankments (Miller and Collins, 1976 and DeLancey, 1976).

Recently, glasphalt has been placed in the city of Baltimore, the city of New York, and the town of Oyster Bay, New York on Long Island (Watson, 1988). The results of survey of this study (Table 2.1) show that four highway agencies, i.e., Connecticut, District of Columbia, New Jersey, and Virginia are considering the use of waste glass as additive to: wearing course (District of Columbia and Virginia), base course (Connecticut, New Jersey, and Virginia), and subbase course (Connecticut). Two states, i.e., Connecticut and Virginia, have recently conducted feasibility studies and issued the reports on the "Feasibility of Utilizing Waste Glass in Pavements" (Larsen, 1989) and "Feasibility of Using Recycled Glass in Asphalt" (Hughes, 1990).

The Connecticut report (Larsen, 1989) investigates the use of waste glass in bituminous pavements and portland cement concrete pavements/structures. The report is based on a comprehensive review of the literature on both laboratory and field evaluations. The Virginia report (Hughes, 1990) is based on laboratory evaluation and economic analysis of glasphalt. It also contains a brief discussion on the use of waste glass in embankment construction and in unbound aggregate layers. Excerpts from these reports are included in the subsequent discussion.

3.2.2 Use of Glass in Asphalt Pavements

The ConnDOT study (Larsen, 1989) reports the following on the technical feasibility and economic aspects of using waste glass in bituminous pavements.

- Glasphalt was successfully mixed and placed in at least 45 locations in the U.S. and Canada between 1969 and 1988. However, most glasphalt has been placed on city streets, driveways and parking lots, and not on high-volume, high-speed highways.
- Potential problems with glasphalt include: loss of adhesion between asphalt and glass; maintenance of an adequate level of skid resistance; and breakage of glass and subsequent raveling under studded tires.
- Glasphalt should be used only as a base course (if laboratory mixes prove acceptable) to minimize potential skid resistance and surface raveling problems.
- Maximum glass size of 3/8 in. should be used in glasphalt, with hydrated lime added to prevent stripping.

- Production of glasphalt would be uneconomical (estimated at $5/ton or 15% more than the conventional hot mix asphalt in Connecticut, under "ideal" conditions).

The limited laboratory study conducted by the Virginia DOT (Hughes, 1990) used two glass contents, i.e. 5% and 15%, and two asphalt contents (based on 50-blow and 75-blow compactive effort) of Virginia S-5 surface mix. The gradation of the basic S-5 mix and recycled glass is given in Table 3.4. The optimum asphalt contents were 6.2% and 5.75% for 50-blow and 75-blow compaction, respectively. The study reports the following trends applicable to asphalt mixes containing glass content of 15% or less:

- the use of glass tends to reduce the voids in mineral aggregate (VMA) and voids in total mix (VTM), and increase voids filled with asphalt (VFA) from Marshall-compacted specimens;
- resilient modulus and tensile strengths are not adversely affected;
- although both wet strength and tensile strength ratio (TSR) moisture damage values were unaffected, some separation at the asphalt/glass interface was observed.

The study indicates that the use of glass in asphalt mixes is technically feasible (with some reservations about the ability of glass to resist moisture damage), if several restrictions are observed. These include: glass content be restricted to 15% or less; the optimum asphalt content must be determined with the target percent of glass to be used; gradation controls are to be 100% passing the No. 3/8 in. sieve and a maximum of 6% passing the No. 200 sieve; and with a TSR of the mix to be 0.9 or higher. On the economic feasibility, (Hughes, 1990) concludes that "there is little monetary incentive to use recycled glass at the present time" in glasphalt in Virginia.

Table 3.4: Gradation of Basic S-5 Mix and Recycled Glass (after Hughes, 1990)

Sieve Size	Gradation of Basic S-5 Mix Without Glass (% Passing)	Gradation of Recycled Glass (% Passing)
1/2 in.	100	100
3/8 in.	95	98
# 4	58	70
# 8	39	32
# 16	29	19
# 30	19	10
# 50	10	6
# 100	6	4
# 200	4.7	2.9

3.2.3 Use of Waste Glass in Portland Cement Concrete

The feasibility study conducted by ConnDOT (Larsen, 1989) concluded that glass is not suitable for placement in portland cement concrete pavement or structures in ConnDOT facilities. The conclusion is mainly based on the study reported by the American Society of Testing and Materials (ASTM; Johnston, 1974) which indicated that glass is highly susceptible to alkali-aggregate reaction (the glass being the reactive aggregate). The reaction between glass and cement causes expansion of glass and reduction in the concrete strength. The elongated particles typical of glass cullet also present a problem with the workability of the concrete mix.

Caltrans (1990) also prohibits the use of glass as an aggregate substitute in portland cement concrete (PCC), cement treated base (CTB), lean concrete base (LCB), and cement treated permeable base (CTPB), due to likelihood of alkali-silica reactions (see Table 2.5).

3.2.4 Use of Glass in Unbound Aggregate Base Layers and Embankment Construction

The use of glass in unbound aggregate base layers is technically feasible (Hughes, 1990). However, the use of glass as an aggregate will require it to be crushed to the appropriate gradation (as per the specifications) and pre-treated if the level of contamination is not within the acceptable limits.

The use of glass as a fill material for embankment construction is preferred to its use in pavements due to the potential problems identified above. However, glass will have to be crushed, although the size will not be critical as in the case of pavements, and the level of contamination of glass will have to be determined for environmental acceptability.

The economic feasibility of the use of glass in unbound aggregate base layers and embankment construction depends on many factors including: the development of resource recovery systems; the availability of market for recycled glass; cost of waste glass, crushing and hauling costs; and the cost of conventional aggregate. In those areas where there is no aggregate shortage, this use of glass seems economically viable only if the crushed and contamination-free glass is available within reasonable distances.

3.2.5 Discussion

Unmarketable glass can be used in highway construction in place of conventional aggregate in asphalt pavement (glasphalt) and in unbound base layers and as a fill material in embankment construction. As previously stated, all these uses have technical, economic, and environmental implications, which must be addressed prior to extensive use of glass in INDOT facilities.

Although glasphalt has been used at a number of locations in the past, long-term performance evaluations have not been conducted (Larsen, 1989), therefore correlations between laboratory test results and field performance are severely lacking. The major areas where potential problems in the use of glasphalt have been identified and further investigation is required include:

- the effect of moisture on glass-asphalt mix;
- type and quantity of the most suitable antistripping agent (presently 1% hydrated lime is used; Larsen, 1989);
- the glass content (Hughes, 1990 recommends 15% to be the upper limit) and most appropriate glass gradation;

- optimum asphalt content and evaluation procedures for asphalt.

The use of waste glass in concrete pavement or structures is not feasible due to alkali-aggregate reaction, and consequent expansion of glass, and reduction in the concrete strength. The problem needs to be further investigated to find the remedial measures.

The use of glass as an aggregate/fill materials in unbound base layers/embankment is feasible if gradation/size meets the INDOT specifications.

It is likely that unmarketable waste glass, available for highway construction, may be contaminated with foreign materials which may include: (a) durable materials (e.g; ceramics, pottery, mirror, pyrex, etc.); (b) nondurable (e.g; wood/metal pieces, cardboard container covers, etc.); or (c) hazardous materials (e.g.; chemically contaminated glass, small batteries, etc.). If the glass contains durable materials, it may be acceptable. However, glass contaminated with the other two categories of materials, depending upon the level of contamination, may require secondary sorting in the case of nondurable materials and pre-treatment (or may even be rejected) in case of hazardous materials. Both secondary recovery and pre-treatment would increase the cost of waste glass. The environmental acceptability of waste glass will depend on the level of contamination, to be determined prior to its use in highway construction.

Based on the results of feasibility studies summarized above, rational conclusions can be drawn regarding the economic feasibility of the use of glass in INDOT facilities. The use of glass in:
- asphalt and concrete pavements will be uneconomical (the cost will be at least 10 to 20% higher than the conventional materials);

- unbound base layers and embankment may be economically justified (however, it will depend on many factors which include: current and projected quantities of recyclable waste glass, crushing and transportation costs, and availability and cost of conventional aggregates);
- highways will reduce landfill costs.

Finally, an adequate and consistent supply of glass is an important factor which will influence its use in the highway industry. The EPA publication on "Characterization of Municipal Solid Waste in the United States: 1990 Update" (EPA, 1990) comments as follows on the production trends of waste glass :

> "Generation of glass has continued to decline from 1986 to 1988. In fact, glass containers would disappear from the waste stream if a trend line analysis were followed. The consultants elected not to use that projection, but to assume that glass containers will continue to be made. The projected generation for 2000 was, however, lowered by 23 percent based on the historical data."

The above comments and the data summarized in Table 3.2 do raise serious concerns about the adequate and consistent supply of recyclable glass.

3.3 Reclaimed Paving Materials

The results of the questionnaire survey of this study show that reclaimed paving materials are the most widely used waste products in a variety of applications by the United States highway industry. Of the 42 state highway agencies which responded to the questionnaire, 41 are engaged in testing, evaluation and use of these materials in subgrade/embankment, subbase, base and wearing courses; see Table 2.2. The experiences of state highway agencies indicate that the use of these materials is economically viable (cost competitive with the virgin materials), technically feasible (performance very good to satisfactory) and generally acceptable from an economic viewpoint (good to satisfactory). Three state highway agencies have expressed their concern over the air pollution from effluents during heating of reclaimed asphalt pavements (RAP), and one state highway agency has identified potential problems with the use of waste concrete in embankments due to its basic chemistry.

3.3.1 Recycling of Asphalt Pavements

Recycling of asphalt pavement is not a new concept. The first mention of recycling is in Warren Brothers portable asphalt plant sales brochure of 1915 (Gannan et al., 1980). However, it was not until the oil crisis of the early 1970's, which rapidly increased asphalt prices and energy costs, that recycling became a feasible method for lowering highway construction costs. In Indiana, the concept of recycling bituminous surface roadway was practiced for the first time by Elkhart County in 1971. Recycled material was used as a base for bituminous overlays and as a subsurface for a chip-and-seal course (Sargent, 1977).

Recycling is generally classified by the type of operation used to perform the recycling. The Asphalt Institute, The Asphalt Recycling and Reclaiming Association (ARRA), the National Asphalt Pavement Association (NAPA) and the U.S. Army Corps of Engineers classify recycling as (Wood et al, 1988):

Hot Mix Recycling - involves removal and mixing at a central plant;
Cold Mix Recycling - may be performed in place or at a crushing plant;
Surface Recycling - is the reworking of one inch of a pavement.

There have been numerous laboratory, field and also synthesis studies on the various aspects of hot mix and cold mix recycling (e.g. NCHRP, 1978; Ferreira et al., 1987; Jordison and Smith,1986; Ganung and Larsen, 1987; Wood et al, 1988, 1989;

Larsen, 1988). NCHRP (1978) also documents the experience of a number of states in recycling asphalt pavements.

The ARRA sponsored a study to develop standard design procedures and specifications for cold in-place recycling of asphalt pavements. Phase A of this study has been accomplished at Purdue University (Wood et al, 1988, 1989). It is based on a literature review and a survey of state and local highway agencies and contractors. The researchers summarized the current practice of cold in-place recycling (CIR) and gave their recommendations to improve mix design, construction and testing procedures. Wood et al (1989) concluded that CIR had shown satisfactory performance and considerable cost savings over conventional overlays.

Iowa

The Iowa DOT has considerable experience in hot mix recycling of asphalt concrete pavements since 1976. Ortgies and Shelquist (1978), Henely (1980), and Jordison and Smith (1986) considered the recycling of asphalt pavements as cost effective, technically feasible and environmentally acceptable. Initially some concern was expressed over air pollution, which was controlled by using an augering device during plant operation. Jordison and Smith (1986) described the hot mix recycling on a demonstration project which consisted of three highway resurfacing projects in Cass and Montgomery Counties. Their experience indicated that:
- recycling asphalt concrete into another highway is a cost effective and non-polluting method of disposal;
- a high quality surface can be constructed using recycled asphalt cement concrete.

Kansas

Maag and Fager (1990) described the experience of the Kansas DOT in hot mix recycling and cold recycling of asphalt pavements. Five different test sections (11.8 miles) were constructed on K-96 in Scott County, Kansas and their performance was monitored for 9 years. The nine-year evaluation concluded that hot and cold recycling (with no additive) are cost effective when compared to a normal overlay.

Recycling of asphalt pavement is a proven fact and many viable processes exist. It is generally cost effective, and recycling of pavements have a positive impact on the environment. The potential problem of air pollution from asphalt plant operation can be reduced by installing emission control devices to make it environmentally safe. There is a need to standardize the design, construction, testing and evaluation procedures. However, the information on recycling of rubber modified asphalt and asphalt-rubber pavements is severely lacking. This issue is discussed in subsection 3.1.4.

3.3.2. Recycling of Concrete Pavements

The recycling of Portland cement concrete pavements has been researched and practiced for a number of years in the United States (e.g., Calvert, 1977; Marks, 1984; Adams, 1988). There are numerous reports issued by various agencies and researchers on different aspects of this technology. Experiences of a few state highway agencies and research findings are briefly described below to identify some of the potential problem areas and to highlight the benefits of recycling cement concrete pavements.

The Iowa DOT has concluded a number of laboratory and field studies on recycling of cement concrete pavements, primarily with a view to reduce the waste disposal problem,

conserve natural resources and achieve monetary saving (Britson and Calvert, 1977; Calvert, 1977; Bergren, 1977; and Marks, 1984). Their experience demonstrated that:
- recycled PCC pavements exhibited good performance;
- the major problem with recycled PCC pavement has been a high frequency of mid-panel cracking; such faulting is typical of conventional pavement without load transfer assemblies;
- there is a substantial economic benefit for recycling PCC pavement in areas where quality aggregate sources are not in close proximity.

Ganung and Larsen (1986) described a six-year evaluation of recycled Portland cement concrete pavement test sections (1000 ft) constructed on Interstate 84 in Waterbury, Connecticut. The nine inch pavement was placed, one 12-ft lane at a time, in three lanes with 40 ft slabs. A 600 ft control section of new concrete was laid adjacent to the above. The salient conclusions of the study included (Ganung and Larsen, 1986):
- energy expenditure for the construction of the control and recycled sections were approximately equal.
- the differences between the two sections were insignificant as far as wear, structural soundness and friction levels were concerned.
- the amount of cracking in the recycled pavement was higher than that of the control, but at that time, no major structural failure had been observed.

Finally, Ganung and Larsen (1986) commented that, "while the installation of the recycled concrete has served as a valuable testbed for future projects, there does not appear to have been any other great advantage to its use in this location".

Arnold (1988) described the experience of recycling concrete pavements by the Michigan DOT. The type of equipment used, procedures, performance and the problems encountered during various phases of recycling operations, i.e. breaking, removal, crushing and paving, were briefly reviewed. Arnold (1988) reported no major difficulty in recycling concrete pavements. However, he suggested that the factors to consider were: the porous and absorptive nature of coarse aggregate; and contamination in the crushed material (e.g., clay balls, bituminous patching materials, joint seals etc.). He recommended that fine aggregate so obtained should not be used in drainage layers. The ENR (1988) reported that "The Michigan DOT figures that recycled aggregate costs about the same as virgin aggregates for concrete pavements. This includes the cost of disposing of old pavements, if new aggregate is used".

Ravindrarajah et al (1987) have reported the results of a laboratory study on using recycled concrete as coarse and fine aggregates in concrete mixes at National University, Singapore. The results of this study are:
- properties of recycled aggregates differed from those of natural aggregates due to the presence of a considerable portion of mortar attached to natural aggregates as well as loose mortar;
- for a medium strength concrete, strength and modulus of elasticity are reduced by about 10% and 30% respectively, whereas drying shrinkage is nearly doubled (after 90 days) when recycled aggregates are used instead of natural aggregates in comparable mixes.

The above studies indicate that recycling of PCC pavements is a technically and economically viable option. In addition, recycling of pavements will have a positive impact on environments, as it will reduce the waste disposal problems. However, further research is needed to address the potential problems (e.g. cracking of recycled concrete pavement), and to refine the mix design procedures and construction techniques. There is also a requirement to more thoroughly consider the economics of recycling PCC pavements, since many factors, which vary with the local conditions, are involved.

3.4 Slags and Ashes

Slags and ashes, derived from the iron, steel and electrical power industries, are perhaps the waste materials of greatest interest to the highway industry, given their wide availability and scope of uses. Large quantities of slags and ashes are currently produced in Indiana (Lovell, 1990). The by-products of the iron and steel industry which have been historically used in the highway industry are iron blast furnace slag and steel slag. The by-products of coal burning plants, which have been widely tested in service and are found useful for a wide range of engineering applications, are coal dry bottom ash, wet bottom ash (wet bottom ash is commonly called boiler slag) and fly ash. The various aspects of use of iron blast furnace slag, steel slag, coal bottom ash, and fly ash are briefly described below.

3.4.1 Iron Blast Furnace Slag

Iron ore, coke and limestone are heated in the blast furnace to produce pig iron. Produced simultaneously in the blast furnace is a material known as blast furnace slag. It is defined as "the non-metallic by-product consisting essentially of silicates and aluminosilicate of lime and other bases", and it leaves from the blast furnace resembling molten lava (Miller and Collins, 1976).

Selective cooling of the liquid slag results in four distinct types of blast furnace slag:(1) air-cooled (solidification under ambient conditions), which finds extensive use in conventional aggregate applications; (2) expanded or foamed (solidified with controlled quantities of water, sometimes with air or steam), which is mainly used as light weight aggregate; (3) granulated (solidified by quick water quenching to a vitrified state), which is mainly used in slag cement manufacture; and (4) pelletized (solidified by water and air-quenching in conjunction with a spinning drum), which is used both as a light weight aggregate and in slag cement manufacture. The bulk of iron blast furnace produced in the United States is of the air cooled variety (Emery, 1982).

Miller and Collins (1976) rank iron blast furnace as having the highest potential among the waste materials for use in highways. Emery (1982) reports that most of the air-cooled blast furnace slag produced in North America is used in granular base, ballast, trench fill and engineering fill. He identified the features of air-cooled blast furnace slag that make it attractive for such applications, and which include: low compacted bulk density (typically 1200 to 1450 kg/m^3) that reduces dead load, lateral pressures, and transportation costs on a volumetric basis; high stability (California Bearing Ratio> 100) and friction angle (approximately 45°); ability to stabilize wet, soft underlying soils at an early construction stage; placeable in almost any weather; very durable with good resistance to weathering and erosion; free draining and non frost susceptible; almost complete absence of settlement after compaction; and non-corrosive to steel and concrete. The leachate does not contain significant concentrations of toxic constituents. He mentioned that a particularly advantageous use of blast furnace slag is in difficult fill conditions over soft ground.

The results of the questionnaire survey of this study show that currently 15 state highway agencies use blast furnace slag in various highway applications (see Table 2.2). It is reported that its use in highways is cost effective, and performance is very good to satisfactory. Nine state highway agencies have reported its environmental acceptability from good to satisfactory, whereas the Illinois DOT identified a potential environmental problem, and the Idaho DOT considers it as environmentally unacceptable.

Limited use of air-cooled blast furnace slag is reported in both dense graded and open graded asphaltic concrete due to high asphaltic cement content requirements (about 8% for dense graded). Air-cooled blast furnace slag is also used as coarse aggregate in all types of concreting operations associated with road construction, i.e.: pavements, precast and prestressed units, foundations, curb and gutter; and ready mix (Emery, 1982).

The engineering properties of air-cooled blast furnace and current practice indicate that the use of air-cooled iron blast furnace slag in a variety of highway applications is economical and technically feasible. Some doubts expressed by state highway agency about its environmental acceptability need to be further investigated.

3.4.2 Steel Slag

Steel slag is a by-product of the steel industry. It is formed as the lime flux reacts with molten iron ore, scrap metal, or other ingredients charged into the steel furnace at melting temperatures around 2800 F. During this process, part of the liquid metal becomes entrapped in the slag. This molten slag flows from the furnace into the pit area where it solidifies, after which it is transferred to cooling ponds. Metallics are removed by magnetic separation (Miller and Collins, 1976).

Steel slags are highly variable, even for the same plant and furnace. Steel slags have high bulk density, and a potential expansive nature (volume change of up to 10% attributed to the hydration of calcium and magnesium oxides). In view of their expansive nature, steel slags are not feasible for use in Portland cement concrete. However, there are many applications where such expansion is tolerable, and has been controlled by suitable aging or treatment (Emery, 1982).

The applications of steel slags in highway construction have been in pavement bases and shoulders, fills, asphalt mixes, and ice control grit. Their most promising application is in asphalt mixes, since asphalt coating eliminates the expansion related problems, and the overall performance of the mixes has been excellent (particularly skid resistance qualities; Emery, 1982).

Nine state highway agencies have reported the use of steel slags in various highway applications (see Table 2.2). Their experience shows that the use of steel slags in highways is generally economical, technically feasible and environmentally acceptable. The expansive nature of the steel slag has been identified as a potential problem. One of the state highway agencies has expressed concern over the leachates, which may be undesirable from environmental viewpoint.

3.4.3 Bottom Ash

The materials collected from the burning of coal at electric utility plants are refered to as power plant ash. These are produced in two forms: bottom ash and fly ash. Bottom ash is the slag which builds up on the heat-absorbing surfaces of the furnace, and which subsequently falls through the furnace bottom to the ash hopper below. Depending upon the boiler type, the ash under the furnace bottom is categorized as dry bottom ash - the ash which is in solid state at the furnace bottom; or wet bottom ash - the ash which is in molten state when it falls in water. It is more often called boiler slag.

Of the 17.5 millions tons of bottom ash produced in 1986 in the United States, 13.4 and 4.1 million tons were dry bottom ash and wet bottom ash, respectively. Only 26.7% of the dry bottom ash was used, whereas 51% of the wet bottom ash was used (ACAA, 1986; reported by Ke et al. 1990). The result of the questionnaire survey of this study show that 7 state highway agencies currently use dry bottom ash and boiler slag in various highway components (see Table 2.2)

The INDOT sponsored a study to determine the feasibility of using bottom ash in highway construction. The study was conducted by two graduate students under the direction of Professor C. W. Lovell at Purdue University (Ke, 1990; Huang, 1990; Huang and Lovell, 1990; Ke et al. 1990). The laboratory studies were performed in three phases: ash characterization tests, engineering properties tests, and environmental evaluation. Extensive laboratory testing was performed on representative samples of Indiana bottom ashes, to determine their physical, chemical and mechanical properties. It was found that the two types of bottom ash have different physical and chemical characteristics and consequently, differently engineering properties (Huang, 1990). The results of the studies (Ke et al., 1990) suggested that bottom ashes have a non-hazardous nature, minimal effects on groundwater quality; low radioactivity, and low erosion potential, but they may be potentially corrosive. They concluded that untreated bottom ashes can be extensively used in highway construction, such as embankments, subgrades, subbases, and even bases. However, those bottom ashes having high corrosiveness should not be placed in the near vicinity of any metal structure.

The economic analysis (Huang, 1990) indicated that while both types of bottom ashes are economical for utilization in Indiana, use of dry bottom ash is more cost effective.

66   Use of Waste Materials in Highway Construction

Huang(1990) recommended that a test section incorporating bottom ash be constructed for monitoring and in situ testing to develop correlation of the laboratory results with the field performance.

3.4.4   Coal Fly Ash

Fly ash is the finely divided residue that results from the combustion of ground or powdered coal and is transported from the combustion chamber by exhaust gases. It is a siliceous material which, in the presence of water, combines with lime to produce a cementitious material with excellent structural properties (see Table 3.5 for typical chemical composition of cement and fly ash). However, the properties depend on the type of coal burning boiler, which are of three types: (1) Stoker Fired Furnaces - not usually good for highway purposes; (2) Cyclone Furnaces - not usually good for use in portland cement concrete and not widely available; and (3) Pulverized Coal Furnace - usually the best quality and quantity and are produced in large quantities (Boles, 1986).

Fly ash represents nearly 75% of all ash wastes generated in the United States (Miller and Collins, 1976). The survey on the current practices in the United States in the use of waste materials in highway construction show that fly ash is the second most widely used waste product in practice. However, there is still much opportunity to expand the use of this product. Table 3.5 shows that in 1984, 80% of the fly ash produced in the United States was wasted in disposal areas.

Use of Fly Ash in Cement Concrete Mixes

The uses of fly ash in the highway industry are included in Table 3.5. The technology for use of fly ash in Portland cement concrete (PCC) and stabilized road bases

Table 3.5: Fly Ash Composition, Production, and Uses (After Boles, 1986)

(a). Typical Chemical Composition

Compounds	Fly Ash Class F	Fly Ash Class C	Portland Cement
SiO_2	54.9	39.9	22.6
Al_2O_3	25.8	16.7	4.3
Fe_2O_3	6.9	5.8	2.4
CaO (lime)	8.7	24.3	64.4
MgO	1.8	4.6	2.1
SO_3	0.6	3.3	2.3

(b). Annual Fly Ash Production*

Production/Disposition	Million Tons	Percentage
Produced	51.3	100
Disposed	40.9	80
Reclaimed	10.4	20

(c). Fly Ash Uses

Uses	Million Tons	% of Total Production
Cement and Concrete Production	5.5	10.7
Structural Fills, Embankments	1.9	3.6
Road Base	0.4	0.7
Filler in Asphalt Mixes	0.1	0.2
Grouting	0.3	0.5
Miscellaneous	2.2	4.3
Total Used	10.4	20.0

* US Production for 1984.

is fairly well developed and has been practiced for many years. There is an abundance of published literature (including synthesis study and technology transfer guidelines; NCHRP, 1976 and Boles, 1986) on the various aspects of the use of fly ash in PCC and stabilized bases. The knowledge that PCC can benefit from the addition of fly ash was recognized as early as 1914. Subsequent research has identified many benefits of the addition of fly ash in concrete mixes, which include: improved workability, which reduces the water requirement - thus resulting in lower bleeding and consequently more durable surface; reduced heat of hydration; increased ultimate strength; increased resistance against alkali aggregates; resistance to sulfate attack (fly ash combines with lime making it less available to react with sulfate); reduced permeability; and economy (Boles, 1986). However, benefits realized will depend on the type of cement, fly ash, mix design and construction procedures.

The Oregon DOT conducted a study on the "Evaluation of Fly Ash as an Admixture in Portland Cement Concrete" (Maloney, 1984), and recommended that the use of fly ash for substructure work, walks, curbs, barriers, other noncritical structures, and cement treated base (CTB) is appropriate as a substitute for Portland cement for a minimum of 10% to a maximum of 20% by weight of Portland cement. The use of fly ash was not recommended for bridge decks, heavily loaded PCC pavements or prestressed concrete in the Oregon DOT facilities. However, a subsequent study (Petrak, 1986), based on the performance of a test section in which fly ash was used in lean concrete base (LCB) and continuously reinforced cement concrete (CRC), concluded that the use of fly ash in these structures was not "technically inappropriate". Requirement of greater curing time was identified as a potential problem.

Use of Fly Ash in Embankments

The fly ash application in which significant quantities can be consumed, with consequent engineering benefits at a much lower cost than conventional materials, is its use in embankments. This application is presently not much practiced. Of the 31 state highway agencies which reported use of fly ash, only 4 are using it as additive to subgrade/embankment and 2 as subgrade/embankment material. Accordingly the implications of such uses are less known, especially from an environmental viewpoint. However, limited use of fly ash in embankments has shown great promise. Faber and DiGioia (1976) described case histories of embankment projects in which fly ash was used as a fill material (3 projects in United States and 3 in England) and the experience is described as successful and economical.

Lewis (1976) has described the construction of a fly ash highway test embankment in Illinois. It is reported that no difficulty was faced during the construction of embankment and the performance of the embankment had been satisfactory with respect to both structural stability and aesthetics. On the environmental aspects, Lewis comments that, "environmental hazards, real or purely speculative, must be solved or fly ash use may never reach its full potential."

Bacher (1990) described the application of fly ash on highway embankments in Delaware (Project ASHRAMP). The embankment (Figure 3.1) served as ramps connecting Interstate 495 with Edge Moor Road and Governor Printz Boulevard. The Electrical Power Research Institute (EPRI) funded the project ASHRAMP to document the design, construction, performance, and environmental characteristics associated with using compacted fly ash for highway embankments. The ASHRAMP project contractor reported that with proper moisture conditions, the placement and resulting performance of the ash ramps were no different than traditional soil material ramps. Two and a half years of sampling and monitoring showed that the fly ash fill had not measurably or detrimentally affected groundwater quality. Bacher (1990) concludes that fly ash is an environmentally and technically acceptable alternative to natural soil when used in an unstabilized form on highway applications.

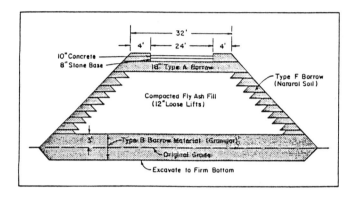

Figure 3.1: Typical Cross Section Showing Fly Ash Fill Surrounded by Borrow Material (from Bacher, 1990).

Martin et al. (1990) investigated the geotechnical properties of ashes from Pennsylvania, Delaware, and New Jersey for use in a highway embankment for Interstate 495 near Wilmington, Delaware. The study included laboratory testing on mechanical and environmental aspects, construction observations, and post construction monitoring. The general conclusions from this study included:
- fly ash must be protected from direct exposure to the atmosphere as it is susceptible to wind and water erosion, and in cold regions, frost heaving;
- fly ash does not generally possess the properties required for use as a highway base course or select fills for slabs on grade.

- the chief value of fly ash in highway and building-site development is as common fill to raise and/or level grades.

The above described field experiences and the results of the survey of this study (Table 2.2) indicate that the use of fly ash in embankments is highly promising. This application results in significant engineering benefits and also ensures disposal of large quantities of fly ash. However, the properties of fly ash vary with the source of coal and the type of plant. Furthermore, groundwater monitoring over a significant period is required to justify firm conclusions about the effects of this application on the groundwater quality.

3.5 Building Rubble

Building rubble discussed in this subsection includes any suitable construction material resulting from the destruction/demolition and removal from any existing structures and buildings. Urban renewal activity may greatly increase such quantities. A large quantities of these recyclable materials are generated annually in the United States, which are mostly landfilled. Miller and Collins (1976) estimated the annual production of building rubble in the United States as 20 million tons, whereas roofing waste constituted 9 million tons (Paulsen et al., 1988). Building rubble is generally a heterogeneous mixture of concrete, plaster, steel, wood, brick, piping, asphalt cement, glass, etc. Paulsen et al. (1988) estimated that roofing waste contains about 36% asphalt cement, 22% hard rock granules (minus No. 10 to plus No. 60 sieve size), 8% filler (minus No. 100 sieve size material) and smaller amounts of coarse aggregate and miscellaneous materials. Substantial variability in the constitution of building rubbles is also expected. However, it is important to consider the feasibility of its use in highway construction, since large quantities of this material may be generated as a result of some catastrophic activity, like earthquakes. Five million tons of concrete debris were generated in the 1989 San Fernando earthquake (Wilson et al. 1976, reported by Ravindrarajah,1987).

The results of the questionnaire survey of this study show that two state highway agencies presently use building rubble as an additive to base and subbase courses, or as a fill material in subgrade/ embankment construction (Table 2.2). Only ConnDOT commented on the performance, stating that the use of building rubble as an additive to base and subbase courses is cost effective and environmentally acceptable.

A study on establishing the feasibility of utilizing roofing waste in asphalt paving mixtures is under way at the University of Nevada. Salient conclusions from the preliminary investigation, based on laboratory testing of samples of roofing wastes from five sources, included (Paulsen et al. 1988):

- acceptable paving mixtures can be produced which contain up to 20% by volume of roofing waste;
- proper selection of binder type and quantity is critical to the performance of the mixture and depends on the type and quantity of the roofing waste in the mixture;
- improved asphalt cement extraction and recovery processes need to be developed to effectively determine the properties of the asphalt cement in the roofing waste;

The ConnDOT conducted a study on the feasibility of expanding the use of demolition materials (ConnDOT, 1988). They defined demolition materials as "any suitable material resulting from the destruction and removal of any existing structure, building, or roadway". It is inferred from the conclusions of their study that building rubble, if processed to make it contamination free, may be used as embankment and subbase course material.

The research and experience in the use of building rubble indicate that it has a potential for use as subbase and subgrade/embankment material. However, its technical and environmental suitability must be determined prior to its use. The economics of using building rubble will depend on many factors, including its feasibility for the various applications (including determining the level of contamination), processing (if required), crushing to appropriate size (depending on the specifications), transportation to the location, and the cost of competing natural aggregate.

3.6 Sewage Sludge

Sewage sludge is created as solids are removed from waste water during treatment. Over the last 20 years, the rate of sludge production has doubled. Currently, over 7 million dry tons of sludge are produced each year in the United States. Most of the sludge is harmless organics. Nutrients like nitrogen and phosphorus are also present and can make for an effective fertilizer. But sludge also contains contaminants taken from wastewater, such as heavy metals, organic carcinogens and pathogens (Morse, 1989).

Sludge disposal is presently regulated by a combination of national and state laws. EPA mandates treatment steps, such as the "process to further reduce pathogens" that are necessary before applying certain sludges to agricultural land (Morse, 1989). Current conditions of sludge disposal in the United States are shown in Figure 3.2. Two disposal methods of sewage sludge i.e., land application and incineration are of interest to the highway industry. Composting facilities have also been built to convert sewage sludge into fertilizer, which can be used for landscaping on highways. The use of sewage sludge for land application in highway operation would be for fertilization of highway medians, which falls in the category of non-agriculture usage and would be governed by the level of sludge contaminants described by the EPA's regulations.

The results of the questionnaire survey of this study are as following:
- Mn/DOT- experimenting with the use of sewage sludge ash, as additive to wearing course and base course. This application is indicated to give satisfactory performance at competitive cost. However, environmental acceptability is doubtful. The Mn/DOT (1990b) recommended that testing must continue for the next two years to verify the suitability- of sludge ash from technical and environmental standpoints.
- New Jersey DOT - use of sludge as compost is economical; its environmental acceptability is marginal.
- Pennsylvania DOT - currently use for landscaping as fertilizer and soil aeration.

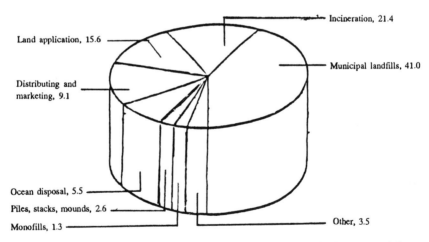

Figure 3.2: Percentages of Sludge Disposal in the U. S., Source is EPA (from Morse, 1989)

3.6.1 Compost and Co-Compost

Caltrans conducted a study on the "Evaluation of Compost and Co-compost Materials for Highway Constructions (Sollenberger, 1987). Compost is derived from the decomposition of sewage sludge with or without a carbonaceous bulking agent. Co-compost is a mixture of sewage sludge or other comparable materials and at least 80% household refuse. Phase I of the study has been completed which contains a literature review and a questionnaire survey. They also distributed the questionnaire to waste facilities in Europe to gain additional information on co-compost, since only one operating co-composting facility currently exists in the United States. Salient conclusions of the study are as follows (Sollenberger, 1987):

- Good quality compost and co-compost products, containing safe and permissible quantities of chemical, biological, and physical contaminants may be used as soil amendments, fertilizers, and erosion control materials with no apparent short term environmental impacts.
- Undesirable impacts may be produced in the environment, such as, heavy metal leachate runoff and a negative aesthetic value, if compost or co-compost materials are used in the construction of sound walls or safety barriers or as embankment material.
- Allergic reactions may occur in individuals who use or apply compost or co-compost materials or who live in the vicinity where these products are used. In very rare instances a common fungus found in these products may cause severe illness or death.
- The long-term health effects caused by exposure to the heavy metals, toxic organics, and pathogenic organisms commonly found in compost and co-compost products are not known.

- Contamination of edible food crops grown on private property adjacent to right-of-way may occur where compost or co-compost products are used.
- Only limited criteria exist pertaining to the quality characteristics of finished compost products. Criteria include pile temperature during the composting process and maximum concentrations of lead, cadmium, and polychlorinated biphenyls (PCBs).
- No explicit regulations, standards, or guidelines were found to exist in the United States pertaining to the use of co-compost products.

3.6.2 Incinerator Ash from Sludge

According to the EPA, there are 282 incinerators burning sludge in the United States (Morse, 1989). The ash residue from the incineration of sewage sludge is nearly completely free of organic matter, is composed almost entirely of silt size material, and contains concentrations of up to 40% lime, which is usually added during dewatering. The ash is not soluble in water but is highly soluble in acid (Miller and Collins, 1976).

Recently Mn/DOT, Minnesota Pollution Control Agency (MPCA), and Metropolitan Waste Control Commission (MWCC) jointly concluded a comprehensive laboratory and field study to assess the feasibility of using sewage sludge ash in bituminous paving (Mn/DOT, 1990b). The sludge ash used in the study was generated as a result of incineration of residual solids from waste water treatment. Three demonstration projects using sludge ash as a fine aggregate in bituminous paving were undertaken. Salient conclusions of the study are as follows:

- Sewage sludge ash has the physical characteristics of conventional mineral fillers and can be fed into bituminous mixtures using equipment similar to that needed for conventional fillers;

- Sludge ash in amount up to 3-4% by weight of the bituminous mixture has potential benefits to bituminous production, including increased strength (resistance to rutting) and reduced asphalt requirements;
- Bituminous mixtures containing 2% sludge ash present no handling, mixing, or paving problems during construction;
- Environmental tests (laboratory leaching of core samples) of bituminous pavement containing 2% sludge ash has shown that all leachate parameters tested were below drinking water standards, except for one parameter on one acid leach test on a single sample.
- Use of sewage sludge as a mineral filler was practical and economically feasible.

Miller and Collin (1976) reports the results of a laboratory examination of the engineering properties of incinerated sewage sludge ash, particularly compaction, compressive strength, freeze-thaw resistance, and age hardening properties. On the basis of test results, it was concluded that incinerated sewage sludge ash possesses many of the properties required for a suitable subbase material.

The current practice and the results of a limited research on the use of sewage sludge in highway operations indicate that the by-products of sewage sludge, i.e., compost and incinerated ash have potential for use as a fertilizer and as an aggregate, respectively. The use of compost is beneficial but has potential safety and environmental risks, whereas use of sewage ash as an aggregate has technical, economical, and environmental implications. Further investigations of sewage sludge by-products are needed to evaluate the risks/repercussions associated with their usage, as compost or as an aggregate in highway construction.

3.7 Incinerator Residue

EPA (1990) estimates show that 25.5 million tons of MSW was burned in 1988 in the United States. It has also been reported that numerous new facilities are scheduled to come into operation in the 1990's. According to projected estimates, 45.5 million tons of MSW will be incinerated in 1995, increasing to 55 million tons in the year 2000 (see Table 3.6). This demonstrates increasing trends in the production of incinerator residue.

Burning MSW results in the formation of two ash products; bottom ash and fly ash. Bottom ash is the unburned and incombustible residue left on the boiler grates after incineration. It consists of slag, glass, rocks, metals, and unbound organic matter, and is composed of large particles (0.1 - 100 mm). Fly ash consists of small-diameter particles of burned or partially burned organic matters, on to which various components of the flue gas have condensed. These particles are usually 1 - 500 μ in size and are entrained in the flue gases. The constituents of fly ash are largely dependent on the type of air pollution control device in use (Blaisdell et al., 1990).

Bottom ash is the larger of the two components, typically comprising 80 - 99% in weight of the total ash (Walsh et al. 1987, reported by Blaisdell, 1990). Typically, MSW incinerated ash amounts to about 25% (dry weight) of the unprocessed MSW input. It was common practice in the past to dispose of incinerator residue along with MSW. However, incinerator ash is no longer classified as MSW (EPA, 1990), and requires separate disposal.

The results of the questionnaire survey of this study show that currently two state highway agencies are experimenting with the use of this waste product in highway construction: (1) the Mn/DOT is evaluating its use as an additive to wearing courses; their experience indicates that the environmental acceptability of the incinerator residue for this

Table 3.6: State of Incineration of Municipal Solid Waste (after EPA, 1990)

Year	Variations from 1960-88							Projected Estimates		
	1960	1965	1970	1975	1980	1985	1988	1995	2000
Generation of MSW, Millions of Tons	87.8	103.4	121.9	128.1	149.6	161.6	179.6	199.8	216.0
Incineration with Energy Recovery, Millions of Tons	-	0.2	0.4	0.7	2.7	7.6	24.5	45.0	55.0
Incineration with Energy Recovery, % of Total Generation	-	0.2	0.4	0.5	1.8	4.7	13.6	-	-
Incineration without Energy Recovery, Millions of Tons	27.0	26.8	24.7	17.8	11.0	4.1	1.0	-	-
Incineration without Energy Recovery, % of Total Generation	30.8	25.9	20.3	13.9	7.4	2.5	0.6	-	-

application is doubtful; (2) the Missouri DOT is experimenting with its use as an aggregate in the base course and also as an additive to wearing courses.

Teague and Ledbetter (1979) investigated the feasibility of using incinerator residue as an aggregate in bituminous base courses. Their study involved the construction and evaluation of approximately 200 ft of roadway on a city street in Houston, Texas. Incinerator residue, from Houston's Holmes Road incinerator plant, was used as an aggregate in a 6 in. thick bituminous base course. This was covered by approximately 1.5 in. of conventional hot-mixed asphalt concrete wearing coarse. A conventional aggregate bituminous base was also constructed adjacent to the test section and evaluated. Three-year performance evaluation has been reported as "extremely well, almost identical with the conventional black base control section." However, it is reported that Hveem stability values for the incinerator residue test section have dropped significantly with time. The cause of this is not known to the investigators.

Tay et al. (1982) conducted a laboratory study at the National University of Singapore to examine the feasibility of utilizing incinerator residue in concrete. They examined the physical and chemical properties of the residue fractions passing a 5 mm (3/16 in.) sieve. Their tests showed that the chemical composition of the various residue samples was fairly constant, whereas the physical properties of the "unwashed" residue varied widely between batches. The use of "raw" residue as a fine aggregate in a typical concrete mix resulted in delayed setting of concrete. However, "washed" samples as a fine aggregate compared favorably with natural sand in terms of compressive strength at ages greater than 7 days.

Public Works (1990b) reports that in Europe, bottom ash from MSW incinerators is used as a base course in secondary road construction.

Patankar et al. (1979 and 1982) conducted a comprehensive study on the economic and environmental factors influencing the use of incinerators residues in bituminous highway construction. They chose five Standard Metropolitan Statistical Areas (SMSA; Chicago, IL; Harrisburg, PA; Miami, FL; New York, NY; and Washington, D.C.) to evaluate the economic and environmental factors of using incinerator residue as opposed to quarried materials. Their analysis was based on the following assumptions.

- 6% hydrated pulverized lime would be added to make the residue have acceptable quality for use as a subbase.
- density of residue ranges from 45 to 50 lb/cf.
- pugmill processing of the residue would be required to screen the material to acceptable highway specification requirements.
- 2% additional asphalt would be used in bituminous mixtures due to greater surface area of the residue compared to conventional aggregates.

The salient conclusions of the study include:

- incinerator residue would prove competitive with virgin aggregate in three of the five metropolitan areas studied (most areas of Chicago, New York and Washington, D.C.; interested readers are advised to study their model to know the exact economic benefits achieved in various parts of the SMSA's studied);
- the use of fused residue in wearing courses was considered uneconomical.

The limited research and experience in the use of incinerator residue as an aggregate indicate that potential for its use in highway construction does exist. However, further research is needed to determine its properties in order to evaluate its feasibility from technical and environmental standpoints. An economic analysis would also be required to justify its use in highway construction.

4. Summary, Conclusions and Recommendations

4.1 Summary

This study synthesizes the information on the use of waste materials in highway construction. The information was obtained from a review of published literature supplemented by: recent unpublished reports, presentations of research updates by professionals at different forums, and personal meetings with the experts. In addition, a questionnaire regarding the use of waste materials was prepared and distributed to each of the state highway agencies. A majority of states responded to the questionnaire, giving a summary of current practices in the use of waste materials in highway construction and their experiences on the technical, environmental, and economic aspects of the various applications of the materials.

Section 1 of this report gives the background, states the objectives of this study, and describes the research approach. Section 2 describes the experience of INDOT in the use of waste materials and summarizes the results of the questionnaire survey. Based on the current practice reported by the respondent state highway agencies, it was considered appropriate to discuss certain waste products in some detail, which included: rubber tires, waste glass, reclaimed paving materials, slags and ashes, building rubble, sewage sludge, and incinerator residue. These waste products are discussed in Section 3 of this report. The conclusions and recommendations of this study are presented in subsequent subsections.

Summary, Conclusions and Recommendations 83

4.2 Conclusions

As a result of this study, the following conclusions are drawn concerning the use of various waste materials in highway construction.

<u>General</u>

- A number of waste products are currently being used (and/or being studied experimentally) in a variety of highway applications by the United States highway agencies. The often used waste products include (in descending order of the number of users who responded to the survey questionnaire of this study), reclaimed paving materials, fly ash, rubber tires, blast furnace slag, steel slag, bottom ash, used motor oil, boiler slag, waste paper, and mine tailings (see Table 2.2). In addition, of the 42 highway agencies which responded to the survey questionnaire of this study, 2 of the state highway agencies also use (and/or study experimentally) building rubble, waste glass, sawdust, ceramic waste, sewage sludge, incinerated residue, and highway hardware.
- Current practice indicates that reclaimed paving materials and slags and ashes are generally used in large quantities.
- Limited available data suggest that reclaimed paving materials, waste tires, and slags and ashes are produced in large quantities in Indiana.

<u>Rubber Tires</u>
- Use of asphalt-rubber as a joint/crack sealant seems cost effective in view of its better performance in most of the cases. However, its long term performance must be monitored due to lack of sufficient experience with its use.

- Use of SAM's reduces the reflection of fatigue cracks of moderate width and thermal cracks; has generally provided longer service life than the conventional surface treatments; and is likely to be equal to the conventional surface treatment on a life cycle cost basis.
- SAMI's have generally not been effective in eliminating the reflection of fatigue cracks; some reduction in reflection of cracks has been experienced, but the performance is not commensurate with the additional cost.
- Asphalt-rubber and rubber modified asphalt in hot mix asphalt (HMA) mixtures have met with both successes and failures. The products need to be further researched to fully understand their behavior prior to their extensive use in the highway industry.
- The initial costs of asphalt paving products with crumb rubber additive (CRA) are generally 50 to more than 100% higher than the products with conventional materials, depending upon the local conditions. The additional costs may be justified over the life cycle, if long-term evaluations show that asphalt-rubber paving products perform better than the conventional materials and provide longer service lives, which is generally not substantiated by the field experience at present.
- The use of CRA in asphalt paving products is generally acceptable from an environmental viewpoint, with some concern about air pollution as a result of adding rubber to the mix and also the requirement of elevated temperatures during mixing of paving materials.
- The use of shredded tires in subgrade/embankment as a lightweight fill material is technically feasible and economically beneficial, as tires are non-biodegradable and large quantities of waste tires can be so consumed. Potential problems include leachate of metals and hydrocarbons. Drinking water Recommended

Allowable Limits (RALs) are found to be exceeded under 'worst-case' conditions (MPCA, 1990).
- The use of tires for soil reinforcement in highway construction is feasible from technical and economic viewpoints, but may have environmental implications.
- The use of tires in retaining structures is economical and practical, but has aesthetic and environmental implications.
- Feasibility of recycling asphalt paving products containing CRA is not known, due to limited reported experience.

<u>Waste Glass</u>

- Glass has been used in the past as an aggregate in asphalt concrete pavements (glasphalt) with some success. Potential problems with glasphalt include: separation at the asphalt/glass interface under moist conditions and subsequent raveling under studded tires; and maintenance of adequate skid resistance.
- Glass content should not be more than 15% (based on the results of reported laboratory study (Hughes, 1990)), if use of glasphalt is considered with hydrated lime to prevent stripping.
- Use of glass in portland cement concrete pavements or structures is not suitable due to the likelihood of alkali-silica reactions.
- The use of glass in subbase and base courses is technically feasible, if it meets the INDOT gradation specifications.
- Glass may be used in embankment construction, if it is crushed to the appropriate size.
- Glass containing non-durable or hazardous materials is not suitable for use in highways; a secondary sorting or pre-treatment may be necessary depending on the level of contamination.

- Consistent and reliable sources of supply of recyclable glass are essential. Current estimates indicate a continuous decline in glass generation (EPA, 1990).
- Economics of the use of glass depend on many factors that vary with the local conditions. The use of glass in asphalt and portland cement concrete pavements is likely to cost more than conventional materials and about the same for unbound base layers and embankment construction. However, there will be a saving in waste disposal costs.

Reclaimed Paving Materials

- The technology for the use of reclaimed paving materials is fairly well developed and it is currently the most widely used waste product in the United States.
- Recycling of asphalt pavements is cost effective and technically feasible. Some concerns are expressed about air pollution in the case of HMA recycling. However, air pollution control devices have been developed to control the emission from asphalt plant within the acceptable limits.
- Recycling of portland cement concrete is technically and economically viable, and results in positive impact on the environments. Potential problems include cracking of recycled concrete pavements.

Slags and Ashes

- The use of iron blast furnace slag as an aggregate in highway construction is technically feasible. Some concerns are expressed about leachates from this waste product.
- The use of steel slag in highway construction as an aggregate is economically viable. Its use in portland cement concrete may not be suitable due to expansion

Summary, Conclusions and Recommendations   87

of steel slag under moist conditions. Its most promising application is as an aggregate in asphalt pavements. Concerns are expressed about its use from an environmental viewpoint.

- The use of Indiana coal bottom ash and boiler slag as an aggregate in highway construction, such as embankments, subgrades, subbases, and even bases is feasible economically and technically, but the material may be potentially corrosive.

- Coal fly ash has been used successfully as a substitute for portland cement, structural fill in embankments, stabilized bases, and filler in asphalt mixes in highway construction. Its properties are variable and depend on the properties of the coal burned and the type of plant. The guidelines on mix design, handling and construction are fairly well developed. Some users have experienced a requirement of longer curing time with some types of fly ash, which leads to many practical construction problems. Some apprehensions exist about use of fly ash in critical structures, such as bridge decks, prestressed concrete, etc.

- Limited experience in the use of coal fly ash as a fill material in embankments indicate that its use in unstabilized form on highway applications is beneficial as large quantities of this product can be consumed. However, long-term research and evaluations are required, especially field monitoring, to determine the effects of this application on groundwater quality.

Building Rubble

- Properties and economics of using building rubble in highway construction depend on the nature and type of source, and also the local conditions. Limited past experience and research indicate that roofing waste is economically and technically feasible for use in asphalt paving mixes. Building rubble has some potential for use as a material in subgrade/embankments and even subbases. However, the product needs to be contamination free for any application in highway construction.

Sewage Sludge

- The use of sewage sludge by-products, i.e., compost and co-compost and incinerator ash, has been attempted in the past in landscaping and as a fine aggregate in highway pavements, respectively. At this point in time both of these applications indicate some potential, economically and technically, but have serious environmental implications. The use of compost and co-compost has also been found to be potentially a health hazard for the individuals who apply these products.

Incinerator Residue

- The use of incinerator residue as an aggregate in bituminous bases and portland cement concrete has been researched in the past but is rarely practiced. Technically there is some promise for use in these applications. The economics of use depend on local conditions and involve significant initial costs to set up a plant for removing unwanted constituents from the ashes. However, its use in some areas has been shown to be economically justified. It may be environmentally unacceptable.

Other Materials

- Currently, used motor oil is recycled and also used as fuel for asphalt plants. Air pollution is a potential environmental concern in the latter use.
- Waste paper is currently being recycled or used as a mulch in landscaping. It may have more economical uses.
- The use of sawdust as a lightweight fill material in highway embankments over soft ground is considered economically and technically feasible. However, its service life is less since it is biodegradable. It gives better performance if used in saturated zone.
- Recycling of highway hardware is currently practised and is feasible.
- Mine tailings have been successfully used for various applications. Their feasibility for use in highway construction depends on the source.

The relative priorities with respect to the number of users (based on the results of the questionnaire survey), and from economical, technical, and environmental viewpoints (based on subjective assessment) are given in Table 4.1.

Table 4.1: Priority of Waste Materials With Respect to Number of Users, Economic, Technical, and Environmental Factors

Priority With Respect to Number of Users	Economics	Technical	Environmental
Reclaimed Asphalt Pavement	Reclaimed Asphalt Pavement	Reclaimed Asphalt Pavement	Reclaimed Asphalt Pavement
Fly Ash	Fly Ash	Blast Furnace Slag	Fly Ash
Rubber Tires	Bottom Ash	Fly Ash	Bottom Ash
Blast Furnace Slag	Boiler Slag	Bottom Ash	Boiler Slag
Steel Slag	Blast Furnace Slag	Boiler Slag	Blast Furnace Slag
Bottom Ash	Steel Slag	Rubber Tires	Steel Slag
Boiler Slag	Incinerator Residue	Steel Slag	Rubber Tires
Sewage Sludge	Waste Glass	Waste Glass	Waste Glass
Waste Glass	Sewage Sludge	Building Rubble	Building Rubble
Building Rubble	Building Rubble	Incinerator Residue	Incinerator Residue
Incinerator Residue	Rubber Tires	Sewage Sludge	Sewage Sludge

Notes: Priorities assigned to the various waste materials with respect to: (a) the number of users is based on the response to the survey questionnaire of this study (see Table 2.2); (b) economics, technical, and environmental are based on subjective assessment of the researchers and are for the usage which is most widely practised (see Table 2.1). However, experience of the state highway agencies as contained in Table 2.1, potential problems identified in the discussion in Section 3, and impact on the environments have been kept in view.

4.3 Recommendations

The following recommendations are offered regarding the use of waste materials by INDOT.

<u>Rubber Tires</u>

- Asphalt-rubber binder as a joint/crack sealant may be used but its long-term performance should be evaluated.
- Tests sections of SAM's may be constructed along with control sections of conventional surface treatments to evaluate and compare their long-term performance.
- A comprehensive analytical, laboratory and field study is warranted prior to using CRA in hot mix asphalt (HMA) mixtures in INDOT facilities. The study should assess the feasibility, recommend the mix design, and specify construction guidelines. It should also assess the service lives and recycling potential of such pavements.
- Rubber tires may be used in subgrade/embankment as a lightweight fill material above the saturated zone. It may be noted that drainage features of the design become very important for this application.
- Tires may be used for soil reinforcement in embankment construction in the unsaturated zone. However, standard designs for this application are needed, as well as an evaluation of adverse effects on the groundwater quality.

Waste Glass

- The use of glass in bituminous pavements and Portland cement concrete pavements or structures is not recommended unless the potential problems identified with its use in these facilities are addressed through laboratory and field evaluations.
- The use of glass in unbound base layers and embankments is recommended, if it can meet INDOT gradation requirements.
- The level of contamination and economics, with respect to the specific source of supply of glass and the location of facility, must be determined to assess its viability prior to its use in highway construction.

Reclaimed Paving Materials

- Recycling of asphalt pavements may be further expanded, as it is found highly beneficial. Specifications may include the requirement of devices to control emission from asphalt plants within the permissible limits. Construction and mix design procedures may be standardized.
- Recycling of concrete pavements is recommended only for experimental purposes, until design and construction procedures are developed for use of this product in INDOT facilities and potential problems are addressed through further research.

Slags and Ashes

- The use of iron blast furnace slag as an aggregate in highway construction has been successful, with some environmental concern, in other states. The

Summary, Conclusions and Recommendations 93

physical, mechanical and chemical properties of Indiana iron blast furnace slags may be determined to assess their feasibility for use in highway construction.

- The use of Indiana coal bottom ashes may be further expanded along with field monitoring and post-construction evaluations to develop correlations of laboratory results with field performance.

- Coal fly ash may be used as a replacement for portland cement in noncritical structures. Their use in critical structures like bridge decks, prestressed concrete etc. is not recommended at this stage. The use of fly ash in embankments may be considered on experimental basis as large quantities of this waste product can be consumed in this application.

Building Rubble

- Caution is urged in the use of building rubble in highway construction. Each source of this waste product must be analyzed for its technical, economical and environmental suitability.

Sewage Sludge

- The use of by-products of sewage sludge, i.e., compost and co-compost and incinerator ash is not recommended at this stage. Further research is needed to evaluate safety and environmental risks involved.

Incinerator Residue

- Incinerator residue has shown some potential for use in subgrade/embankment, subbase, and base, but experience and research in its use is very limited.

Feasibility studies are recommended to assess technical, economic and environmental implications of its use in INDOT facilities.

Other Materials

- Existing practice of recycling used motor oil should continue. Possibility for its use as fuel in asphalt plants may be considered. It may be possible to control its adverse effects on air quality by taking remedial measures.
- Use of waste paper is recommended for recycling and as mulch in landscaping.
- Sawdust may be used as a lightweight fill material in highway embankments in saturated zones.
- Recycling of highway hardware may be further expanded.
- Use of mine tailings, depending on the type and nature of material and the local conditions, may be considered.

Table 4.2 presents a plan for INDOT for further expanding the use of waste products in highway construction and also identifies potential problems/research areas.

Summary, Conclusions and Recommendations 95

Table 4.2: Recommended Plan for INDOT for the Use of Waste Materials in Highway Construction

Waste Materials	Recommended Priority of Applications	Potential Problems/Research Areas
Reclaimed Paving Material	(1) Recycling of asphalt pavements; (2) Recycling of portland cement concrete (PCC) pavements.	(1) Is cost effective and economically feasible, air pollution is a potential problem; (2) Recycled PCC pavements have experienced cracking, which need to be further researched.
Coal Fly Ash	(1) Additive to PCC pavements; (2) As a fill material in embankment.	(1) Guidelines and procedures are fairly developed, only recommended for noncritical pavement structures, further research is required for use in PCC slabs, bridge decks etc.; (2) Limited experience in this application, need to be further researched to determine properties of Indiana ashes and its effects on environments.
Blast Furnace Slag	As an aggregate in highway construction.	Is technically and economically feasible, properties vary with type of source, some concern over leachates, properties of Indiana slags and effects on environments need to be determined.
Coal Bottom Ash/ Boiler Slag	As an aggregate in highway construction.	Is technically and economically feasible, but may be potentially corrosive, field evaluations are required to develop correlations of laboratory results with field performance.
Steel Slag	As an aggregate in highway construction.	Is technically and economically feasible in asphalt pavements, potentially expansive when exposed to moisture, leachate analyses to determine environmental acceptability are required.
Rubber Tires	(1) Crack/joint sealant; (2) As Stress Absorbing Membrane (SAM); (3) Soil reinforcement; (4) As lightweight aggregate in embankments; (5) As Stress Absorbing Membrane Interlayer (SAMI); (6) As overlays/wearing courses in asphalt pavements.	(1) Has generally performed better than conventional materials, long-term performance need to be evaluated; (2) Reduces reflection of cracking, costs more than the conventional surface treatments, has generally given longer service life, further evaluation is required to determine performance and cost effectiveness over life cycle; (3) Is beneficial as large quantities of tires can be consumed and results in considerable savings in earthfill, leachates have adverse effects on environments; recommended for use only in unsaturated zone, further evaluations are required to develop standard designs and ascertain environmental consequences and economic benefits; (4) Beneficial since nonbiodegradable, leachate problem, recommended to be used in unsaturated zone; (5) Reduces reflection of cracking, is uneconomical; (6) Experience indicates both failures and successes, costs 50% to more than 100% higher than conventional materials, longer life has generally not been substantiated by field performance, concern over air pollution, further research is required to assess performance, environmental acceptability, develop mix design specifications and construction procedures, and feasibility for recycling.
Waste Glass	(1) As an aggregate in unbound base layers and as a fill material in embankments; (2) As an aggregate in asphalt pavements; (3) As an aggregate in PCC pavements.	(1) Contamination-free glass is technically feasible if it meets INDOT gradation specifications, economics depend on local conditions; (2) Is uneconomical, potential problems include: separation at the asphalt/glass interface and subsequent raveling under studded tires, and maintenance of adequate skid resistance; (3) Is not suitable in PCC pavements due to likelihood of alkali-silica reactions.
Incinerator Residue	As an aggregate in highway construction.	Limited experience/research, leachates may contaminate groundwater, further laboratory research is needed to assess economic, technical, and environmental feasibility.
Building Rubble	(1) Use of roofing waste in asphalt pavements; (2) As an aggregate in highway construction.	(1) Is technically feasible, if the source is contamination-free, economics depend on the local conditions; (2) Potentially suitable as a fill material in subgrade/embankment and even in subbases, if it is contamination-free.
Sewage Sludge	(1) Compost and co-compost; (2) Incinerator ash from sludge as an aggregate in highway construction.	(1) Likelihood of some health risk to individuals applying this products and also concern over undesirable leachates; (2) Limited experience, is economical in some areas, environmental concerns have been expressed over leachates, further research is required to assess its feasibility.
Waste Paper*	(1) Recycling; (2) As mulch for landscaping.	Both applications are feasible, recycling is more beneficial.
Used Motor Oil*	(1) Recycling; (2) Fuel in asphalt plants.	(1) Is beneficial; (2) Environmental concern over air pollution.
Highway Hardware*	Recycling.	Is economically beneficially.
Mine Tailings*	As an aggregate in highway construction.	Feasibility depends on source which need to be evaluated from economic, technical, environmental factors.
Sawdust*	As a lightweigh fill material in embankments.	Biodegradable, service life may be less, gives better performance if used in saturated zones as it is likely to reduce decay of wood.

Notes: The priority of applications for various waste materials given in this Table is based on subjective assessment of the researchers. * Not listed according to priority.

5. References

Notes:

ASCE	=	American Society of Civil Engineers
ASTM	=	American Society for Testing and Materials
DOT	=	Department of Transportation
ENR	=	Engineering News Record
FHWA	=	Federal Highway Administration
NAE	=	National Academy of Engineering
NAS	=	National Academy of Sciences
NCHRP	=	National Cooperative Highway Research Program
NRC	=	National Research Council
TRR	=	Transportation Research Record
TRB	=	Transportation Research Board

5.1 References Cited

Adams, B. (1988), "Ohio Sand, Gravel Producer Recycles Broken Concrete," Pit and Quarry, Vol. 80, No. 2.

Allen, H.S. and Turgeon, C. M. (1990), "Evaluation of 'PlusRide™ (A Rubber Modified Plant Mixed Bituminous Surface Mixture) - Final Report," Report No. 90-01, Minnesota Department of Transportation, Mapplewood, MN, 35pp.

American Coal Ash Association, Inc. (ACAA) (1986), Ash Production and Utilization, Washington, D. C.

Anderson, K. W. (1990), "Comments on the Current Practice in Response to the Survey Questionnaire ," Washington State DOT, (Unpublished)

Arm-R-Shield™ (1986), Sales Literature, Spring, 1986.

Arnold, C. J. (1988), "Recycling Concrete Pavements," Concrete Construction, Vol. 33, No., 3, pp. 320-325.

Bacher, J. R. (1990), "Ash is Cash: Fly Ash Applications," Public Works, Vol 120, No. 9, pp. 44-45.

Bergren, J.V. (1977), " Portland Cement Concrete Utilizing Recycled Pavement," Report NO. MLR 77-3, Iowa DOT, 29p.

Bernard, D (1990), "The Issue of Scrap Tires and Their Potential Use as an Additive for Asphalt Paving Products," Statement of Chief Demonstration Project Division, USDOT, Before the Subcommittee on Energy Regulation and Conservation Committee on Energy and Natural Resources United States Senate, Unpublished, 8pp.

Bjorklund, A. (1979), "Proceedings XVI World Road Congress, Vienna, Austria.

Blaisdell, M., Lee, D., and Baetz, B., (1990), " Economic Feasibility of Drying Municipal Solid-Waste Combustion Residue," Journal of Energy Engineering, ASCE, Energy Division, Vol. 116, NO. 2, pp. 87-97.

Boles, W. F. (1986), "Fly Ash Facts for Highway Engineers," Technology Transfer Report FHWA-DP-59-8, Federal Highway Administration, U. S. DOT, Demonstration Projects Division, Washington, D.C., 47 pp.

Breakspere, R. J., et. al. (1980), "New Developments in Waste Glass Utilization," Conservation and Recycling, Vol. 3, pp. 233-248.

Breuhaus, J. H. (1990), " Tire Grinding Systems," Letter of the President of T^3 Technology Limited to the Indiana Department of Environmental Management, October 10, 1990, (Unpublished), 3 pp.

Britson, R., and Calvert, G. (1977), "Recycling of Portland Cement Concrete Roads in Iowa," Report NO. MLR 77-6, Iowa DOT, 8 pp.

Byrum, D. (1971), "Asphalt Pavements from Glass and Rubber Wastes," Rural and Urban Roads, Vol 9, No. 12, pp.24-25.

Caltrans (1985), "Proposed Standard Plan: Reinforced Soil Utilizing Salvaged Guardrails Components," Memorandum Office of Transportation Laboratory, California DOT, Dated Jan. 16, 1985," Sacramento, CA,(Unpublished).

Caltrans (1986), "Alternative Earth Retaining Systems in California Highway Practice," California State DOT, Sacramento, CA, (Unpublished).

Caltrans (1988), "Use of Discarded Tires in Highway Maintenance," Translab Design Information Brochure No. TL/REC/1/88, Sacramento, CA, (Unpublished).

Caltrans (1990), "Caltrans Response to AB 1306,"Memorandum of Caltrans, Office of Transportation Materials and Research, Dated October 15, 1990", Sacramento, CA, (Unpublished) 4 PP.

Calvert, G. (1977), " Iowa DOT's Experience with Recycling Portland Cement Pavement and Asphalt Cement Pavement," Report No. MLR 77-4, Iowa DOT, 12 pp.

Chandler, W. U. (1986), "Materials Recycling: The Virtue of Necessity," Conservation and Recycling, Vol. 9, No. 1, pp. 87-109.

Charles, R. et al. (1980), "Recycling Conventional and Rubberized Bituminous Concrete Pavements Using Recycling Agents and Virgin Asphalt Modifiers (A Laboratory and Field Study)," Asphalt Paving Technology, Proceedings Association of Asphalt Technologist, Vol. 49, Louisville, Kentucky, pp. 95-122.

Colony, D. C. (1979), "Industrial Waste Products in Pavements: Potential for Energy Conservation," TRR, No. 734, pp.16-21.

ConnDOT (1988), " Report to the General Assembly on the Feasibility of Expanding the Use of Demolition Materials in Projects Undertaken by the Department of Transportation," Report NO. 343-20-88-13, Connecticut DOT, Bureau of Highways Office of Research and Materials, Wethersfield, Connecticut, 39 p.

Diamond, S. (1985), " Selection and Use of Fly Ash for Highway Concrete," Report JHRP-85-8, Purdue University, West Lafayette, IN 47907.

Emery, J. J. (1982), "Slag Utilization in Pavement Construction," Extending Aggregate Resources, ASTM STP 774, ASTM, pp. 95-118.

ENR (1972), "Pavement in Half Glass and Concrete Waste," Engineering News Record, Vol. 189, No. 17, pp.

ENR (1980), "Rubber in Asphalt May Cut Skidding," ENR, vol. 204, No. 18, p.24.

ENR (1982), "Sawdust Fill Fights Slides," ENR, p. 25.

ENR (1983), "Road Use for Waste By-products," ENR, pp. 13-14.

ENR (1988), "Highways: The Second Time Around," ENR, pp. 38-39.

EPA (1990), " Characterization of Municipal Solid Waste in the United State: 1990 Update," Report No. EPA/530-SW-()-042, United States Environmental Protection Agency, 103 pp.

Epps, J. A. et al. (1980), "Guidelines for Recycling Asphalt Pavements," Asphalt Paving Technology, Proceedings Association of Asphalt Technologist, Vol. 49, Louisville, Kentucky.

Esch, D. C. (1984), "Asphalt Pavements Modified with Coarse Rubber Particles-Design, Construction, and Ice Control Observations," Report No. FHWA-AK-RD-85-07, Alaska DOT AND Public Facilities, Fairbanks, AK, 35pp.

Faber, J. H. and DiGioia, A. M. (1976), "Use of Ash in Embankment Construction," TRR, No. 593, pp. 13-19.

Ferreira, M. A., Servas, V. P., and Marsais, C. P. (1987), "Accelerated Testing of Recycled Asphalt Concrete," Proceedings Association of Asphalt Paving Technologists, Vol. 56, Technical Session, Reno, Nevada, pp. 259-277.

Forsyth, R. A., and Egan, J. P. (1976), "Use of Waste Materials in Embankment Construction," TRR, No. 593, pp. 3-8.

Gannon, C. R. et al. (1980), "Recycling Conventional and Rubberized Bituminous Concrete Pavements Using Recycling Agents (A Laboratory and Field Study)," Asphalt Paving Technology, Proceedings Association of Asphalt Technologist, Vol. 49, Louisville, Kentucky, pp. 95-122.

Ganung, G. A., and Larsen, D. A. (1986), " Portland Cement Concrete Pavement Recycling, I-84, Waterbury - Final Report," Report No. FHWA-CT-RD646-F-86-14, Connecticut DOT, Wethersfield, CN.

Ganung, G. A., and Larsen, D. A., (1987), "Performance Evaluation of a Hot-Mixed Recycled Bituminous Pavement Route 4, Burlington - Final Report," Report No. FHWA-CT-RD-647-4-87-1, Connecticut DOT, Wethersfield, 40pp.

Halstead, W. J. (1986), "Use of Fly Ash in Concrete (Final Report)," Report No. TRB/NCHRP/SYN-127, Transportation Research Center, Washington, D. C., 74pp.

Heine, M. (1990), "Recycling `Romance' Makes RAP's Future Bright," Roads and Bridges, Vol. 28, No. 10, p. 32.

Henely, R. P. (1980), " Evaluation of Recycled Asphalt Concrete Pavements - Final Report," Research Project No. HR-1008, Kossuth County, Iowa, 24pp.

Henely, R. P., and Schiek, R. (1982), " Recycled Asphalt Pavement - Final Report," Project No. HR176, Iowa Highway Research Board, 22pp.

Huang, W. H. (1990), "The Use of Bottom Ash in Highway Embankment and Pavement Construction," Doctor of Philosophy Thesis, Purdue University, West Lafayette, IN, 317pp.

Huang, W. H. and C. W. Lovell, (1990), "Bottom Ash as Embankment Material," Geotechnics of Waste Fills - Theory and Practice, ASTM STP 1070, pp. 71-85.

Hughes, C. S., (1990), "Feasibility of Using Recycled Glass in Asphalt," Report No. VTRC90-R3, Virginia Transport Research Council, Charlottesville, Virginia, 21pp.

Huls, J. M. (1989),"Managing Wastes in the Caribbean," Biocycle, Vol 30, No. 7, pp. 40-42.

Humprey, T. (1990), " Cold In-Place Recycling Low-Cost Option State," Roads and Bridges, Vol. 28, No. 10, pp. 48.

Hunsucker, D. Q. et al. (1987), "Road Base Construction Utilizing Coal Waste Materials," Report No. UKTRP-87-15, Kentucky Research Program, College of Engineering, University of Kentucky, Lexington, Kentucky, 16pp.

Hunsucker, D. Q. (1990), " Questionnaire on Indiana Department of Highway's Synthesis Study; Use of Waste Materials in Highway Construction," Memorandum No. H.2.117 in Response to Survey Questionnaire of This Study, Dated Nov. 28, 1990 (Unpublished).

Hunsucker, D. Q., and Sharpe, G. W. (1989), " The Use of Fly Ash in Highway Construction," Research Report No. KTC-89-32, Kentucky Transportation Center, College of Engineering, University of Kentucky, 41pp.

Hunsucker, D. Q., and Graves, R. C. (1989), " Preliminary Engineering, Monitoring of Construction, Initial Performance, and Evaluation - Use of Ponded Fly Ash in Highway Road Bases," Research Registration No. KTC-89-56, Kentucky Transportation Center, College of Engineering, University of Kentucky, 47pp.

IDEM (1991), "Indiana Tire Stockpiles," Indiana Department of Environmental Management, Tire Stockpile List Dated March 15, 1991, 6 pp.

INDOT (1988), Indiana Department of Transportation Specifications, Section No. 403.04B6.

Johnston, C. D. (1974), "Waste Glass as a Coarse Aggregate for Concrete," Journal of Testing and Evaluation, ASTM, Vol. 2, No. 5, pp 334-350.

Jordison, D., and Smith, R. D. (1986), "Asphalt Cement Concrete Pavement recycling - Cass and Montgomery County: Final Report," Project No. HR-1018, Iowa DOT, 53pp.

JAWMA (1990), " Scrap Tire Disposal Problem to be Addressed By New Council," Journal of the Air and Waste Management Association, Vol. 40, No. 6, p. 841.

Kandhal, P. (1990), " Asphalt Answers to Recycling Questions," Roads and Bridges, Vol. 28, No. 10, p. 16.

KDOT (1990), "Economics of Using Asphalt Rubber in Pavements," Kansas DOT, (Unpublished).

Ke, T.-C.(1990), "The Physical Durability and Electrical Resistivity of Indiana Bottom Ash," Research Report No. FHWA/IN/JHRP-9076, School of Engineering, Purdue University, West Lafayette, IN, 335pp.

Ke. T.-C., Lovell, C. W., Huang, W. H. and Lovell, J. E. (1990), "Bottom Ash as a Highway Material," Draft Paper for 70th Annual Meeting of the Transportation Research Board January, 1991," (Unpublished), 38pp.

Kekwick, S. V. (1986), "A Bituminous -Rubber Pavement Rehabilitation Experimentation Using the Heavy Vehicle Simulator," Asphalt Paving Technology, Proceedings Associations of Technologists, Vol. ,Clear Water Beach, Florida, pp. 400-418.

Keller, G. R. (1990), "Retaining Forest Roads," Civil Engineering, ASCE, Vol. 60, No. 12, pp. 50-53.

Keller, J. J. (1972), " An Experiment With Glasphalt in Burnbay, B.C., (Canada)," Proceedings Canadian Technical Asphalt Association, Vol. 17.

Kilpatrick, D. (1985), "Tyres Help Holderness to Tread New Ground in Coastal Protection," Surveyor, Vol. 165, No. 4875, pp. 10-11.

Larsen, D. A. (1988), " Performance Evaluation of a Cold In-Place Recycled Bituminous Pavement," Route 66, Marlborough - Final Report," Report No. FHWA-CT-RD.647-6-88-1, Wethersfield, 48pp.

Larsen, A. D. (1989), "Feasibility of Utilizing Waste Glass in Pavements," Report No. 243-21-89-6, Connecticut Department of Transportation, Wethersfield, CT, 17pp..

Larsen, D. A. (1989a), "Eight-Year Performance Evaluation of an Asphalt-Rubber Hot Mix Pavement," CT Report No. 116-3-89-8, Connecticut Department of Transportation, Drawer A Wethersfield, CT, 20pp.

Lauer, K. R. (1979), "Potential Use of Incinerator Residue as Aggregate for Portland Cement Concrete," TRR, No. 734, NAS, Washington, D. C., pp. 44-46.

Lewis, T. S. (1976), "Construction of Fly Ash Road Embankment in Illinois," TRR, No. 593, pp.20-23.

Lovell, C. W. (1990), "Personal Communication".

Lundy, J. R., Hicks, R. G., Richardson, E. (1987), "Evaluation of Rubber-Modified Asphalt Performance-Mt. St. Helens Project," Asphalt Paving Technology, Proceedings Associations of Technologists, Vol. 56,, pp. 573-598..

Lucas, D. W. (1990), " Use of Recycled or Waste Products by the Indiana Department of Transportation," A Report to the Indiana Department of Environmental Management, (Unpublished), 5pp.

Maag, R. G., and Fager, G. A. (1990), "Hot and Cold Recycling of K-96 Scott County," Kansas DOT,Division of Operations, Bureau of Materials and Research, Topeka, Kansas, 23pp.

Maag, R. G., and Parcells, W. H., (1982), "Hot Mix Recycling Gray County, Kansas - Final Report," Report No. FHWA-KS-82/3, Kansas DOT, Bureau of Materials and Research, Topeka, Kansas, 67pp.

Maine DOT (1990), "The Use of Tire Rubber in Pavements- Preliminary Report," Maine Department of Transportation, Technical Services Division, Research and Development Section, Maine, 15pp.

Maloney, M. J. (1984), "Evaluation of Fly Ash Admixture in Portland Cement Concrete," Oregon State Highway Division, Oregon, 65pp.

Mark, C. R., et. al. (1972), "Promising Replacements for Conventional Aggregates for Highway Use," NCHRP, Synthesis of Highway Practice Report No. 135, NRC, NAS-NAE, HRB, Washington,D.C., 53pp.

Marks, V. J. (1984), " Recycled Portland Cement Concrete Pavement in Iowa - Final Report," Project No. HR-506, Iowa DOT, 24pp.

Martin, J. P., et al. (1990), " Properties and Use of Fly Ashes for Embankments," Journal of Energy Engineering, ASCE, Energy Division, Vol. 116, No. 2, pp. 71-85.

McCarthy, B. (1990), "What is New in Asphalt Materials?," Better Roads, Vol. 60, No. 1, pp. 24-25.

McDaniel, R. S. (1990), "Personal Communication".

McManis, K. L. and Arman, A. (1989), "Class C Fly Ash as a Full or Partial Replacement for Portland Cement or Lime," TRR, No. 1289, pp. 68-82.

McQuillen, J. L. and Hicks, R. G. (1987), "Construction of Rubber Modified Asphalt Pavements," Journal of Construction Engineering and Management, Vol. 113, No. 4, pp. 537-553.

McQuillen, J. L., Takallou, H. B., Hicks, R. G. and Esch D. (1988), "Economic Analysis of Rubber-Modified Asphalt Mixes," Journal of Transportation Engineering, Vol. 114, No. 3, pp. 259-275.

McReynolds, R. L. (1990), "Current and New Uses of Waste Products in Kansas Highways," Presented at MVC Research Sub Committee (Unpublished)

Mellott, D. B. (1989), "Discarded Tires in Highway Construction (Demonstration Project No. 37; Final Report)," Repot No. FHWA/PA - 89/009+79-02 Pennsylvania Department of Transportation, of Bridge and Roadway Technology, Harrisburg, PA, 15pp.

Miller, R. H. and Collins, R. J. (1976), "Waste Materials as Potential Replacements for Highway Aggregates," NCHRP, Report No. 166, TRB, NRC, Washington, D.C., 94pp.

Mn/DOT (1990), Update: Minnesota Department of Transportation's Commitment to Waste Product Utilization," (Unpublished), 4pp.

Mn/DOT (1990a), "Update On: Willard Munger Recreational Trail Waste Tire and Shingle Scrap Asphalt Paving Test Section," Minnesota Department of Transportation (Unpublished), 3pp.

Mn/DOT (1990b), " Sewage Sludge Ash Use in Bituminous Paving," Report to Legislation Commission on Waste Management, Prepared Jointly by Minnesota DOT, Minnesota Pollution Control Agency, Metropolitan Waste Control Commission. 45 pp.

Morris, G.R. and McDonald C. H. (1976), "Asphalt-Rubber Stress-Absorbing Membranes Field Performance and State of the Art," TRR 595, pp. 52-58.

Morse, D. (1989), "Sludge in the Nineties,"Civil Engineering, ASCE, Vol. 59, No. 8, pp. 47-50.

MPCA (1990), "Waste Tires in Sub-grade Road Beds- A Report on the Environmental Study of the Use of Shredded Waste Tires for Roadway Sub-grade Support," Minnesota Pollution Control Agency, , St. Paul, MN, 34pp.

NCHRP (1972), "Pavement Rehabilitation- Materials and Techniques," NCHRP No. 9, National Research Board, National Research Council, NAS-NAE, Washington, D. C., 41pp.

NCHRP (1976), "Lime-Fly Ash- Stabilized Bases and Subbases," NCHRP No. 37, Transportation Research Board, National Research Council, Washington, D. C., 66pp.

NCHRP (1978), "Recycling Materials for Highways," NCHRP, Synthesis of Highway Practice Report No. 54, TRB, NRC, Washington, D.C., 53pp.

NYSDOT, (1990), "Use of Scrap Rubber in Asphalt Pavements-A Report to the Governor and Legislature in Compliance with Chapter 599 of the Laws of 1987," New York State Department of Transportation, (Unpublished), NYDOT, Albany, NY, 12pp.

ODOT (1990), "Research Notes, Investigation of Raveling on PlusRide™ Section," Oregon DOT, (Unpublished)

Ortgies, B. H., and Shelquist, R. A. (1978), "Recycling of Asphalt Concrete: from I-80 in Cass County - 1011," Iowa DOT, Highway Division, Office of Materials, 27pp.

Patankar, U. M. et. al. (1979), "Evaluation of the Economic and Environmental Feasibility of Using Fused and Unfused Incinerator Residue in Highway Construction, (Final Report, 1978-79)," Report No. FHWA-RD-79-83, JACA Corporation, Fort Washington, PA, 131pp.

Patankar, U. M., Taylor, M. R. and Ormsby, W. C. (1982), "Economics of Using Incinerator Residue as a Highway Construction Material," Extending Aggregate Resources, ASTM STP 774, ASTM. pp. 43-63.

Paulsen, G., Stroup-Gardiner, M., and Epps, J. (1988), " Roofing Waste in Asphalt Paving Mixtures," Center for Construction Materials Research, Department of Civil Engineering, University of Nevada, (Unpublished), Reno, Nevada, 44pp.

Petrak, A. (1986), "Fly Ash in Continuous Reinforced Pavements and Lean Concrete Bases,"Experimental Features Interim Report No. OR 85-01, Oregon State Highway Division, Research Section, Salem, Oregon, 16pp.

PlusRide™ (1984), "PlusRide™ Asphalt-Rubberized Road Surface Compound," PlusRide™ Sales Literature 1984.

Powell, J. (1983), "A Comparison of the Energy Savings from the Use of Secondary Materials," Conservation and Recycling, Vol. 6, No. 1/2, pp. 27-32.

Prendergast, J. (1989), "The Battle Over Burning," Civil Engineering, ASCE, Vol. 59, No. 7, pp 48-51.

Public Works(1990), "Recycling-An Overview," Public Works, Vol. 120, No. 8, pp.53-56.

Public Works (1990a), "Tires," Public Works, Vol. 121, No.4, pp. 80-94.

Public Works (1990b), "Incineration of Solid Municipal Waste: A State of the Art Report," Public Works, Vol. 120, No. 8, pp. 48-50.

Public Works (1990c), "Tire Fill Stabalizes Roadway, Embankment," October, 1990, pp.68.

Ravindrarajah, R. S. (1987), "Utilization of Waste Concrete for New Construction," Conservation and Recycling, Vol. 10, No. 2/3, pp. 69-74.

Ravindrarajah, R. S., Loo, Y. H. and Tam, C. T. (1987), "Recycled Concrete as Fine and Coarse Aggregates in Concrete," Magazine of Concrete Research, Vol. 39, No. 141, pp.214-220.

Sargent, H. J. (1977), "Recycling 50 Miles of Bituminous Pavements Saves Dollars - Expands Road Program," Proceedings of 63 Annual Road School, Purdue University, West Lafayette, IN, pp. 198-209.

Schnormeier, R. H. (1986), "Fifteen-Year Pavement Condition History of Asphalt-Rubber Membranes in Phoenix, Arizona," TRR, Vol. 1096, pp. 62-67.

Seals, R. K., Moulton, L. K., and Ruth, B. E. (1972), " Bottom Ash: An Engineering Material," JSMFED, ASCE, Vol. 98, No. SM4, pp. 311-325.

Simanski, R. E. (1979), "Asphalt Recycling is Here to Stay," Information Series, No. 172, The Asphalt Institute, 3pp.

Singh, J. and Athay, L. D. (1983), "Technical, Cost and Institutional Aspects of Asphalt-Rubber Use as a Paving Material," Conservation and Recycling, Vol. 1, No. 1/2, pp. 21-26.

Sollenberger, D. A. (1987), "Evaluation of Compost and Co-Compost Materials for Highway Construction, Phase 1 (Final Report)," Report No. FHWA/CA/TL-87/04, Caltrans, Sacramento, CA, 97pp.

Stephens, J. E. (1989), "Nine-Year Evaluation of Recycled Rubber in Roads -Final Report," Report No. JHR 89-183, Civil Engineering department, School of Engineering, University of Connecticut, Storrs, CT, 16pp.

Takallou, H. B. and Hicks, R. G. (1988), "Development of Improved Mix and Construction Guidelines for Rubber-Modified Asphalt Pavements," TRR, No. 1171, pp.113-120.

Takallou, H. B., McQuillen, J. and Hicks, R. G. (1985), "Effect of Mix Ingredients on Performance of Rubber-Modified Asphalt Mixtures," Report No. FHWA-AK-RD-86-05, Alaska DOT AND Public Facilities, Fairbanks, AK, 160pp.

Takallou, H. B., Hicks, R. G. and Esch, D. C. (1986), "Effect of Mix Ingredients on the Behavior of Rubber-Modified Asphalt Mixtures," TRR No. 1096, pp. 68-80.

Takallou, H. B., Hicks, R. G. and Takallou, M. B. (1989), "Use of Rubber Modified Asphalt for Snow and Ice Control," Proceedings International Conference Strategic Highway Research Program and Traffic Safety on Two Continents, Gothenburg, Sweden, pp-51-68.

Takallou, M. B., Layton, R. D., and Hicks, R. G. (1987), " Evaluation of Alternative Surfacing for Forest Roads," Transportation Research Record, No. 1106, pp. 10-22.

Tay, J., Tam, C. and Chin, K. (1982), "Utilization of Incinerator Residue in Concrete, Conservation and Recycling, Vol. 5, No. 2/3, pp. 107-112.

Teague, D. J. and Ledbetter, W. B. (1979), "Performance of Incinerator Residue in a Bituminous Base," TRR, No. 734, TRB, NAS, Washington, D. C. pp. 32-37.

TNR, (1985), "Tire - Anchored Timber Walls - Economical and Practical," TNR, Vol. 117, Washington, D. C.

Turgeon, C. M. (1989), "The Use of Asphalt-Rubber Products in Minnesota," Report No. 89-06, Minnesota Department of Transportation,MN, 11pp.

Vallerga, B. A. et al. (1980), "Application of Asphalt Rubber-Membranes in Reducing Reflection Cracking," Asphalt Paving Technology, Proceedings Association of Asphalt Technologist, Vol. 49, Louisville, Kentucky, pp. 330-353.

Vermont Agency of Transportation (1988), "Research Update: Cost Effectiveness of Asphalt-Rubber Surface Treatment on I-91 Springfield-Weathersfield," No. 88-8, Unpublished, 2pp.

Walton, E. C. (1971), "Glasphalt - Borough of Scarborough, Proceedings 5th Annual Meeting of APWA, Ontario Chapter.

Walsh, P., O' Leary, P., and Cross, F.(1987),"Residue Disposal from Waste-to-Energy Facilities," Waste Age, May, pp. 57-65.

Watson, J. (1988), "When the Tire Hits the Glasphalt," Resource Recycling Magazine, pp. 18-21.

Wayne, D. D. (1976), "Waste Use in Highway Construction," TRR, No. 593, pp. 9-12.

Wilson, D. G. et al. (1976), "Demolition Debries - Quantities, Composition and Possibility for Recycling," Proceedings 5th Mineral Waste Utilization Symposium, Chicago, pp. 8-16.

Wood, L. E., White, T. D., and Nelson, T. B. (1988), "Cold In-Place Recycling (CIR)," Department of Civil Engineering, Purdue University, West Lafayette, IN, 69pp.

Wood, L. E., White, T. D. and Nelson, T. B. (19 89), "Current Practice of Cold In-Place Recycling of Asphalt Pavements," Transportation Research Record, Vol. 1178, pp. 31-37.

5.2 Other References

Anderson, D. A. and Dukatz, E. L. (1980), "Asphalt Properties and Composition: 1950-1980," Asphalt Paving Technology, Proceedings Association of Asphalt Technologist, Vol. 49, Louisville, Kentucky, pp. 1-29.

Anderson, D. A. and Dukatz, E. L. (1980), "Asphalt Properties and Composition: 1950-1980," Asphalt Paving Technology, Proceedings Association of Asphalt Technologist, Vol. 49, Louisville, Kentucky, pp. 1-29.

Anderson, M. et al. (1985), " Laboratory Evaluation of Stabilized Flue Gas Desulfurization Sludge (Scrubble Sludge) and Aggregate Mixtures," Research Report No. VKTRP-85-1, Kentucky Transportation Research Program, College of Engineering University of Kentucky, Lexington, Kentucky 166pp.

Baer, A. J. (1989), "Japan's Recycling Empire: Recovery from A to Z," World Wastes, Vol. 32, No. 12, pp. 22-24.

Baladi, G. Y., Ronald, D. H., and Lyles, W. R. (988), "New Relationships Between Structural Properties and Asphalt Mix Parameters," TRR, Vol. 1171, pp. 168-177.

Baykal, G., Arman, A., and Ferrel, R. (1989), " Accelerated Curing of Fly Ash-Lime Soil Mixtures," TRR, NO. 1219, pp. 82-92.

Beardsley, D. (1985), "The Impact of Recycling on the Environment," Conservation and Recycling, Vol. 8 No. 3/4, pp.387-391.

Better Roads (1987), "Asphalt-Rubber Tested as Recycling Binder," , Better Roads, Vol. 57, No. 11, pp. 32-33.

Better Roads (1990), " Recycling Paver Strengthens Maintenance Plan," Better Roads, Vol. 60, NO. 9, pp. 27-28.

Biddulph, M. W. (1976), "Principles of Recycling Processes," Conservation and Recycling, Vol. 1, pp. 31-54.

Biddulph, M. W. (1977), "Cryogenic Embrittlement of Rubber," Conservation and Recycling, Vol. 9, No. 1, pp.169-178.

Bilbrey, J. H., Sterner, J. W. and Valdez, E. G. (1979), "Resource Recovery from Automobile Shredder Residues," Conservation and Recycling, Vol. 2, pp. 219-232.

Bonnaure, F. Gravois, A and Udron, J. (1980), "A New Method for Predicting the Fatigue Life of Bituminous Mixes," Asphalt Paving Technology, Proceedings Association of Asphalt Technologist, Vol. 49, Louisville, Kentucky, pp.499-526.

Brown, S.F. and Cooper, K. E. (1980), "A Fundamental Study of the Stress-Strain Characteristics of a Bituminous Material," Asphalt Paving Technology, Proceedings Association of Asphalt Technologist, Vol. 49, Louisville, Kentucky, pp. 476-497.

Buekens, A. G. (1977), "Some Observations on the Recycling of Plastics and Rubber," Conservation and Recycling, Vol. 1, pp. 247-271.

Button, J. W., Little, D. N., Kim, Y, and Ahmed, J. (1987), "Mechanistic Evaluation of Selected Asphalt Additives," Asphalt Paving Technology, Proceedings Associations of Technologists, Vol. 56, Reno, Nevada, pp. 62-90.

Carrasquillo, P. M., Tikalsky, P. J. and Carrasquillo, R. L. (1986), "Mix Proportioning of Concrete Containing Fly Ash for Highway Applications (Final Report)," Report No. FHWA/TX-87/4+364-4F, Texas University at Austin, Center for Transportation Research, 55pp.

Castedo, H. (1987), "Significance of Various Factors in the Recycling of Asphalt Pavements on Secondary Roads," Transportation Research Record, Vol. 1115, pp. 125-133.

Clark, T. (1985), "Scrap Tyres: Energy for the Asking," British Plastics and Rubber, Mar. 1985, pp. 35-37.

Coetzee, N. E. (1979), "An Analytical Study of the Applicability of a Rubber Asphalt Membrane to Minimize Reflection Cracking in Asphalt Concrete Overlay Pavements," Presented at 59th Annual Meeting, Transportation Research Board.

Collin, R. J. (1979), "Composition and Characteristics of Municipal Incinerator Residues, TRR, No. 734, NAS, Washington, D. C., pp. 46-53.

Collivignarelli, C., Riganti, V. and Urbini, G. (1986), "Battery Lead Recycling and Environmental Pollution Hazards," Conservation and Recycling, Vol. 10, No. 2/3, pp. 177-184.

Custovic, E., et. al. (1987), "Copper Recovery from Secondary Materials in the Shaft Furnace with Used Automobile-Tire Additions," Conservation and Recycling, Vol. 10, No. 2/3, pp. 93-98.

EPA (1981), "Data Collection and Analyses Pertaining to EPA's Development of Guidelines for Procurement of Highway Construction Products Containing Recovered Materials," Report No. EPA/SW/MS-2096-Vol. 1, 1981, Issues and Technical Summaries, Franklin Associates Ltd., Prairie Village, KS, Sponsored by Environmental Protection Agency, Washington, D.C., 195pp.

Deckson, P. F. (1973), "Cold Weather Paving With Glasphalt," Symposium on Secondary Uses of Waste Glass.

Dukatz, A. L. and Anderson, D. A. (1980), "The Effect of Various Fillers on the Mechanical Behavior of Asphalt and Asphaltic Concrete," Asphalt Paving Technology, Proceedings Association of Asphalt Technologist, Vol. 49, Louisville, Kentucky, pp. 530-549.

Emery, J. J. (1978), "Utilization of Wastes and Byproducts as Construction Materials in Canada," Conservation and Recycling, Vol. 2, pp. 31-41.

Galler, S. R. (1985), "The Government Perspective," Conservation and Recycling, Vol. 8, No. 3/4, pp. 441-443

Gotoh, S. (1980), "Source Separation for Resource Recovery - State-of-the-Art," Conservation and Recycling, Vol. 3, pp. 305-317.

Halstead, W. J. (1979), "Potential for Utilizing Industrial Wastes and By-Products in Construction of Transportation Facilities in Virginia - Part I," Report No. FHWA/VA-80/15, Virginia Highway and Research Council, Charlottesville, VA, 78pp.

Harrison, K., Tong, S. G. and Hilyard, N. C. (1986), "An Economic Evaluation of Cryogenic-Grinding of Scrap Automotive Tyres," Conservation and Recycling, Vol. 9, No. 1, pp. 1-14.

Hawkins, W. L. (1987), "Recycling of Polymers," Conservation and Recycling, Vol. 10, No. 1, pp. 15-19.

Henstock, M. E. (1976), "New Products from Waste," Conservation and Recycling, Vol. 1, pp. 161-166.

Henstock, M. E. (1976), "The Scope for Materials Recycling," Conservation and Recycling, Vol. 1, pp. 3-17.

Herz, D. E., Gold, S. M. and Bell, C. O. (1987), "War on Disposables," Biocycle, Vol. 28, Nov-Dec 1987, pp. 46-52.

Hopkins, T. C., Hunsucker, D., and Sharpe, G. W. (1988), " Highway Field Trials of Chemically Stabilized Subgrades, Research Report Dated Oct. 1988, Kentucky Transportation Center, College of Engineering, University of Kentucky, Lexington, Kentucky, pp. 42 pp.

Howell, R. B. and Baumeister, K. L. (1978), "Analysis of Solid Waste Materials in Highway Operations," Report No. FHWA-78-27, California State Department of Transportation, Sacramento, CA, 79pp.

Iwai, S., Takatsuki, H. and Urabe, S. (1980), "The Producing Mechanism, Separative and Fuel Characteristics of Municipal Refuse," Conservation and Recycling, Vol. 3, pp.249-257.

Jimenez, R. A. (1980), "A Look at the Art of Asphaltic Mixture Design," Asphalt Paving Technology, Proceedings Association of Asphalt Technologist, Vol. 49, Louisville, Kentucky, pp.323-352.

Kahn, W. (1985), "Supply/Demand and Pricing of Scrap Materials," Conservation and Recycling, Vol. 8, No. 3/4, pp. 393-398.

Khan, M. Z. A. and Burney, F. A. (1989), "Forecasting Solid Waste Composition - An Important Consideration in Resource Recovery and Recycling," Resources. Conservation and Recycling, Vol. 3, pp. 1-17.

Kaufer, H. (1987), "Active Recycling of Plastics," Conservation and Recycling, Vol. 10, No. 2/3, pp. 153-167.

King, G. N. et al. (1986), "Polymer Modification: Binder's Effect on Mix Properties," Asphalt Paving Technology, Proceedings Associations of Technologists, Vol. ,Clear Water Beach, Florida, pp. 520-537.

Korzum, E. A. (1990), " Economic Value of Municipal Solid Waste," Journal of Energy Engineering, Energy Division, ASCE, Vol. 116, No. 1, pp. 39-50.

Koutsky, J., Clark, G. and Klotz (1977), "The Use of Recycle Tyre Rubber Particles for Oil Spill Recovery," Conservation and Recycling, Vol. 1, pp. 231-234.

Kreiter, B. G. (1976), "Composition of Wastes and Some Possibilities for Recovery," Conservation and Recycling, Vol. 1, pp. 19-30.

Kruse, C. G. (1989), " A Proposal for Testing And Evaluation of Mixed Municipal Refuse Incinerator Ash," Submitted to Hemepin County by Braun Environmental Laboratories, (Unpublished), 18pp.

Lagrone, B. D. (1986), "Reclaiming of Elastomers," Conservation and Recycling, Vol. 9, No. 4, pp. 359-361.

Little, D. N. and Epps, J. A. (1980), "Evaluation of Certain Structural Characteristics of Recycled Pavement Materials," Asphalt Paving Technology, Proceedings Association of Asphalt Technologist, Vol. 49, Louisville, Kentucky, pp. 218-251.

Little, D. N. (1986), "An Evaluation of Asphalt Additives to Reduce Permanat Deformation and Cracking in Asphalt Pavements: A Brief Synopsis of On-Going Research," Asphalt Paving Technology, Proceedings Associations of Technologists, Vol. ,Clear Water Beach, Florida, pp. 314-322.

Malisch, W. R. , Day, D. F. and Wixson, B. G. (1970), "Use of Domestic Glass for Urban Paving -Summary Report," U. S. Environmental Protection Agency, Report No. EPA-670/2-75-035, and Highway Research Record No. 30, Highway Research Board, Washington, D. C.

Malisch, W. R. et al. (1970), "Glasphalt: New Paving Material Completes First Canadian Trial," Engineering and Contract Record.

Milgrom, J. (1980), "Recycling Plastics: Current Status," Conservation and Recycling, Vol. 3, pp. 327-335.

Ming, C (1980), "Organizational Techniques in Retrieving Waste Materials," Conservation and Recycling, Vol. 3, pp. 319-321.

Ming, C (1980), "The Recycling of Waste Acid, Oil and Metal Scrap of the Shanghai Donghai Oil and Chemical Recycling Works," Conservation and Recycling, Vol. 3, pp. 323-325.

Nutting, J. (1990), " New Materials, Resources and Environments," Metals and Materials, Vol. 6, No. 2, pp. 83-86.

Pamukcu, S., Tuncan, M., and Fang, H. Y. (1990), "Influence of Environmental Activities on Physical and Mechanical Behavior of Clays," Physico-Chemical Aspects of Soil and Related Materials, ASTM STP 1095, K. B. Hoddinott, and R.O. Lamb, Eds., ASTM, Philadelphia, pp. 91-107.

Purcell, A. H. (1978), "Tire Recycling: Research Trends and Needs," Conservation and Recycling, Vol. 2, No. 2, pp. 137-143.

Ramachandran, V. S., (1981), "Waste and By-Products as Concrete Aggregates," Canadian Building Digest, No. 215, National Research Council of Canada.

Robinson, S. (1989), "Wave Goodbye to Scrap Tyres," European Rubber Journal, Vol. 171, Sep. 1989, pp. 32-33.

Rwebangira, T., Hick, R. G. and Truebe, M. (1987), "Determination of Pavement Layer Structural Properties for Aggregate-Surfaced Roads," TRR, Vol. 1106, pp. 215-221.

Schiek, R., and Jones, K. (1982), " Evaluation of Control Devices for Asphalt Pavement Recycling Operations - Summary," Report No. HR-188, Iowa DOT, 8pp.

Schmidt, H. (1987), "Economy Considerations in Building Material Recycling," Conservation and Recycling, Vol. 10, No. 2/3, pp. 59-67.

Sherwood, P. T. (1975), "The Use of Waste and Low-Grade Materials in Road Construction: 2. Colliery Shale," Transportation and Road Research Laboratory, Department of the Environment, Material Division, Highway Department, Crowthorne, Berkshire, 18pp.

Siemon, D. M., Agnew, J.B. and Salusinszky, A. L. (1979), "Recycling of Used Lubricating Oil in Australia," Conservation and Recycling, Vol. 3, pp. 25-33.

Spaak, A. (1985), "Recycling a Mixture of Plastics: A Challenge in Today's Environment," Conservation and Recycling, Vol. 8, No. 3/4, pp. 419-428.

Sultan, H. A. (1979), "Copper Mill Tailings, Incinerator Residue, Low Quality Aggregate Characteristics, and Energy Savings in Construction, TRR, No. 734, TRB, NAS, Washington, D. C. 53pp.

Sussman, W. A. (1976), "Reclaimed Glass Aggregate Asphalt Pavement," Highway Focus, Vol 8, No. 2..

Terrel, R. L. and Walter, K. L. (1986), "Modified Asphalt Pavement Materials-the European Experience," Asphalt Paving Technology, Proceedings Associations of Technologists, Vol. Clear Water Beach, Florida, pp. 482-518.

Wisconson DOT (1987), "Fly Ash Embankment," Report on the plans and the Specifications, Project I. D. # 4010-3-00 (Unpublished), Sheboygan County, WI, 16pp.

Whalley, L. (1982), "The Effect of Changing Patterns of Materials Consumption on the Recycling of Cars and Domestic Appliances in the UK," Conservation and Recycling, Vol. 5, No. 2/3, pp. 93-106.

Wuff, S. T. (1985), "The Future of Recycling: The ISIS Perspective," Conservation and Recycling, Vol. 8, No. 3/4, pp. 429-432.

Wutz, M. J. (1987), "Plastic Material Recycling as Part of Scrap Vehicle Utilization - Possibilities and Problems," Conservation and Recycling, Vol. 10, No. 2/3, pp. 177-184.

Yip, W., and Tay, J. (1990), "Aggregate made from Incinerated Sludge Residues," <u>Journal of Materials in Civil Engineering</u>, ASCE, Materials Engineering Division, Vol. 2, No. 2, pp. 84-93.

Yoder, E. J. and Witzak, M. W. (1975), "<u>Principal of Pavement Design</u>," 2nd Ed., John Willey and Sons, Inc. pp 285-286.

Appendix A: Survey Questionnaire

QUESTIONNAIRE
Synthesis Study
USE OF WASTE MATERIALS IN HIGHWAY
CONSTRUCTION

The Indiana Department of Transportation (INDOT) has undertaken a project to search out and synthesize the useful knowledge from all possible sources and to prepare a report on current practices and potential uses of waste materials in highway construction. Your answers to the under mentioned questions will not only help INDOT, but will allow the entire highway fraternity to benefit from your valuable experience/research in the subject areas.

(1) What waste materials do you currently use in highway construction?

—Rubber Tires —Waste Glass —Ceramics Wastes
—Plastics Wastes —Battery Casings —Used Motor Oil
—Sewage Sludge —Incinerator Residue —Pyrolysis Residue
—Coal Fly Ash —Coal Bottom Ash —Boiler Slag
—Blast Furnace Slag —Steel Slag —Foundry Wastes
—Waste Paper —Building Rubble —Reclaimed Paving Material
(Specify)_____
—Mineral Wastes (Specify)_____

__Others (Specify) _____

(2) How is each of the materials marked in (1) used?

    (a) Additive to wearing course. (b) Additive to base course.
    (c) Additive to subbase course. (d) Additive to subgrade/embankment course.
    (e) As a wearing course. (f) As a base course.
    (g) As a subbase course. (h) As a subgrade/embankment course.
    (i) For landscaping. (j) Others (specify)

Materials	Uses (Give Letters)	Materials	Uses (Give Letters)

(3) In what annual quantities (volumes/weights) are the waste materials marked in (1) used?

Materials	Quantities	Materials	Quantities

(4) Are you required by state law to use waste materials? Which ones? What use?

Materials	Usage	Materials	Usage

(5) What is your experience in the use of waste materials with respect to the conditions listed below?

Materials	Cost Effectiveness	Performance	Environmental Acceptability

(6) What materials and uses seem favorable and are projected for future construction?

Materials	Usage	Materials	Usage

(7) Do you use the waste products in a process covered by patents? Which ones?

Materials	Uses	Materials	Uses

Note: You may attach additional sheets, if required for clarity of response. It will by highly appreciated if copies of recent research findings/performance evaluation studies on the use of waste materials in highway construction are returned along with this completed questionnaire at the following address:

Professor C. W. "Bill" Lovell, Room G245, Department of Civil Engineering, Purdue University, West Lafayette, IN 47907.
Phone No. 317 - 494 - 5034   Fax No. 317 - 494 - 0395

Name _____ Title _____
Address _____
Phone No. _____ Fax No. _____
Date _____

List of Abbreviations

ARRA	Asphalt Recycling and Reclaimed Association
ASTM	American Society for Testing and Materials
Caltrans	California State Department of Transportation
CIR	Cold In-Place Recycling
ConnDOT	Connecticut Department of Transportation
CRA	Crumb Rubber Additive
Cu	Cubic
DOT	Department of Transportation
EPA	Environmental Protection Agency
Ft	Foot/Feet
HMA	Hot Mix Asphalt
INDOT	Indiana Department of Transportation
Mn/DOT	Minnesota Department of Transportation
MPCA	Minnesota Pollution Control Agency
MSW	Municipal Solid Waste
NYSDOT	New York State Department of Transportation
PCC	Portland Cement Concrete
PennDOT	Pennsylvania Department of Transportation
RAL	Recommended Allowable Limits
RAP	Reclaimed Asphalt Pavement
RCP	Reclaimed Concrete Pavement
SAM	Stress Absorbing Membrane
SAMI	Stress Absorbing Membrane Interlayer
SMSA	Standard Metropolitan Statistical Area
TCT	Twin City Testing Corporation
US	United States
Yd	Yard(s)

Note: Abbreviations used in the tables are described under each table.

Other Noyes Publications

HANDBOOK OF POLLUTION CONTROL PROCESSES

Edited by

Robert Noyes

This handbook presents a comprehensive and thorough overview of state-of-the-art technology for pollution control processes. It will be of interest to those engineers, consultants, educators, architects, planners, government officials, industry executives, attorneys, students and others concerned with solving environmental problems.

The pollution control processes are organized into chapters by broad **problem areas**; and appropriate technology for decontamination, destruction, isolation, etc. for each problem area is presented. Since many of these technologies are useful for more than one problem area, a specific technology may be included in more than one chapter, modified to suit the specific considerations involved.

The pollution control processes described are those that are actively in use today, as well as those innovative and emerging processes that have good future potential. An important feature of the book is that **advantages** and **disadvantages** of many processes are cited. Also, in many cases, **regulatory-driven trends** are discussed, which will impact the technology used in the future.

Where pertinent, regulations are discussed that relate to the technology under consideration. Regulations are continually evolving, frequently requiring modified or new treatment technologies. This should be borne in mind by those pursuing solutions to environmental problems.

Innovative and emerging technologies are also discussed; it is important to consider these new processes carefully, due to increasingly tighter regulatory restrictions, and possibly lower costs. For some pollutants specific treatment methods may be required; however for other pollutants, specific treatment levels must be obtained.

CONTENTS

1. REGULATORY OVERVIEW
2. INORGANIC AIR EMISSIONS
3. VOLATILE ORGANIC COMPOUND EMISSIONS
4. MUNICIPAL SOLID WASTE INCINERATION
5. HAZARDOUS WASTE INCINERATION
6. INDOOR AIR QUALITY CONTROL
7. DUST COLLECTION
8. INDUSTRIAL LIQUID WASTE STREAMS
9. METAL AND CYANIDE BEARING WASTE STREAMS
10. RADIOACTIVE WASTE MANAGEMENT
11. MEDICAL WASTE HANDLING AND DISPOSAL
12. HAZARDOUS CHEMICAL SPILL CLEANUP
13. REMEDIATION OF HAZARDOUS WASTE SITES
14. HAZARDOUS WASTE LANDFILLS
15. IN SITU TREATMENT OF HAZARDOUS WASTE SITES
16. GROUND WATER REMEDIATION
17. DRINKING WATER TREATMENT
18. PUBLICLY OWNED TREATMENT WORKS
19. MUNICIPAL SOLID WASTE LANDFILLS
20. BARRIERS TO NEW TECHNOLOGIES
21. COSTS

INDEX

In summary, a vast number of pollution control processes and process systems are discussed.

ISBN 0-8155-1290-2 (1991)   7" x 10"   750 pages

ASHEVILLE-BUNCOMBE
TECHNICAL COMMUNITY COLLEGE

3 3312 00043 6972

TE 200 .A45 1993

Ahmed, Imtiaz.

Use of waste materials in
highway construction

DISCARDED

DEC - 6 2024